Key Issues in Environmental Change

Series Editors:

Co-ordinating Editor

John A. Matthews

Department of Geography, University of Wales Swansea, UK

Editors

Ray S. Bradley

Department of Geosciences, University of Massachusetts, Amherst, USA

Neil Roberts

Department of Geography, University of Plymouth, UK

Martin A. J. Williams

Mawson Graduate Centre for Environmental Studies, University of Adelaide, Australia

Preface to the series

The study of environmental change is a major growth area of interdisciplinary science. Indeed, the intensity of current scientific activity in the field of environmental change may be viewed as the emergence of a new area of 'big science' alongside such recognized fields as nuclear physics, astronomy and biotechnology. The science of environmental change is fundamental science on a grand scale: rather different from nuclear physics but nevertheless no less important as a field of knowledge, and probably of more significance in terms of the continuing success of human societies in their occupation of the Earth's surface.

The need to establish the pattern and causes of recent climatic changes, to which human activities have contributed, is the main force behind the increasing scientific interest in environmental change. Only during the past few decades have the scale, intensity and permanence of human impacts on the environment been recognized and begun to be understood. A mere 5000 years ago, in the mid-Holocene, non-local human impacts were more or less negligible even on vegetation and soils. Today, however, pollutants have been detected in the Earth's most remote regions, and environmental processes, including those of the atmosphere and oceans, are being affected at a global scale.

Natural environmental change has, however, occurred throughout Earth's history. Large-scale natural events as abrupt as those associated with human environmental impacts are known to have occurred in the past. The future course of natural environmental change may in some cases exacerbate human-induced change; in other cases, such changes may neutralize the human effects. It is essential, therefore, to view current and future environmental changes, like global warming, in the context of the broader perspective of the past. This linking theme provides the distinctive focus of the series and is mentioned explicitly in many of the titles listed overleaf.

It is intended that each book in the series will be an authoritative, scholarly and accessible synthesis that will become known for advancing the conceptual framework of studies in environmental change. In particular we hope that each book will inform advanced undergraduates and be an inspiration to young research workers. To this end, all the invited authors are experts in their respective fields and are active at the research frontier. They are, moreover, broadly representative of the interdisciplinary and international nature of environmental change research today. Thus, the series as a whole aims to cover all the themes normally considered as key issues in environmental change even though individual books may take a particular viewpoint or approach.

John A. Matthews (Co-ordinating Editor)

Forthcoming titles in the series

Atmospheric Pollution: an Environmental Change Perspective (Sarah Metcalfe, Edinburgh University, Scotland)

Biodiversity: an Environmental Change Perspective (Peter Gel, Adelaide University, Australia)

Climatic Change: a Palaeoenvironmental Perspective (Cari Mock, University of South Carolina, USA)

Cultural Landscapes and Environmental Change (Lesley Head, Wollongong University, Australia)

Environmental Change at High Latitudes: a Palaeoecological Perspective (Atte Korhola and Reinhard Pienitz, Helsinki University, Finland & Laval University, Québec, Canada)

Environmental Change in Drylands (David Thomas, Sheffield University, UK)

Environmental Change in Mountains and Uplands (Martin Beniston, Fribourg University, Switzerland)

Natural Hazards and Environmental Change (W.J. McGuire, C.R.J. Kilburn and M.A. Saunders, University College London, UK)

Pollution of Lakes and Rivers: a Palaeoecological Perspective (John Smol, Kingston University, Canada)

The Oceans and Environmental Change (Alastair Dawson, Coventry University, UK)

Wetlands and Environmental Change (Paul Glaser, Minnesota University, USA)

Glaciers and Environmental Change

Atle Nesje

Department of Geology, University of Bergen, Norway

and

Svein Olaf Dahl

Department of Geography, University of Bergen, Norway

Hodder Arnold

A MEMBER OF THE HODDER HEADLINE GROUP

First published in Great Britain in 2000 by
Hodder Arnold, an imprint of Hodder Education and a member of
the Hodder Headline Group, an Hachette Livre UK company,
338 Euston Road, London NW1 3BH

www.hoddereducation.com

© 2000 Atle Nesje and Svein Olaf Dahl

British Library Cataloguing in Publication Data
A catalogue record for this book is available from the British Library

Library of Congress Cataloging-in-Publication Data
A catalog record for this book is available from the Library of Congress

ISBN 978 0 340 70634 3

Production Editor: Wendy Rooke
Production Controller: Iain McWilliams
Cover Design: Mouse Mat Design

Typeset in 10/11$\frac{1}{2}$ Palatino by Academic and Technical, Bristol

What do you think about this book? Or any other Hodder Arnold title?
Please send your comments to educationenquiries@hodder.co.uk

Contents

	Preface	ix
	Acknowledgements	xiii
Chapter 1	**Introduction**	1
	1.0 Chapter summary	1
	1.1 The significance of environmental change	1
	1.2 Glaciers as monitors of environmental change	2
	1.3 The study of glaciers and past glacier fluctuations in the context of present and future environmental change	6
Chapter 2	**Theories for climate and glacier variations**	9
	2.0 Chapter summary	9
	2.1 The astronomical (Milankovitch) theory of climate variation	9
	2.2 Variation in atmospheric gas content and climate change	11
	2.3 Volcanic activity and climate variations	13
	2.4 Variations in solar output	14
	2.5 Geodynamic factors	15
Chapter 3	**Methods of palaeoenvironmental reconstruction**	17
	3.0 Chapter summary	17
	3.1 Stable isotope variations in ice cores	17
	3.2 Climate records from ice cores	22
	3.3 Gas content in ice cores	31
	3.4 Microparticles and radioactive tracers in ice cores	35
	3.5 Volcanism and climate	37
	3.6 Mapping and measuring glacier-front variations	38
	3.7 Terminal moraines	39
	3.8 Lacustrine sediments	43
	3.9 Marine sediments	45
Chapter 4	**Glacier dynamics**	48
	4.0 Chapter summary	48
	4.1 Present distribution of glaciers	48
	4.2 Glacier types	48
	4.3 Temperature distribution in glaciers and ice sheets	53

4.4 Glacier monitoring	56
4.5 Glacier monitoring by satellites	56
4.6 Determination of the equilibrium line altitude (ELA)	57
4.7 Reconstruction of the equilibrium line altitude	58
4.8 Mass balance	61
4.9 Frontal variations	92
4.10 Response time/time lag	93
4.11 Glacier movement	98
4.12 Supraglacial ice morphology	107
4.13 Glacier hydrology	108
4.14 Calving glaciers	111
4.15 Surging and tidewater glaciers	112
4.16 Reconstruction of ice-surface profiles and calculation of basal shear stress	113

Chapter 5	**Glacier variations**	**116**
	5.0 Chapter summary	116
	5.1 Pre-Quaternary glaciations	116
	5.2 Glacial/interglacial cycles during the Quaternary	119
	5.3 Late Cenozoic glacier and climate variations	122
	5.4 Late-glacial glacier and climate variations in NW Europe	135
	5.5 Variations of local glaciers during the last glaciation	137
	5.6 Holocene glacier and climate variations	138
	5.7 Neoglacial glacier variations	144
	5.8 Little Ice Age glacier variations	144
	5.9 Glaciers, environmental change and the human race	159
	5.10 Models of Late Quaternary climate and ice-sheet evolution	160

Chapter 6	**Late Quaternary sea-level changes**	**162**
	6.0 Chapter summary	162
	6.1 Glacio-eustasy and glacio-isostasy	162
	6.2 Relative sea-level changes	164
	6.3 Sea ice	168
	6.4 Models of future sea-level changes	171

Chapter 7	**Models of future climatic change**	**174**
	7.0 Chapter summary	174
	7.1 Orbitally-induced (Milankovitch) climate change	174
	7.2 Greenhouse gas-induced warming	175
	7.3 Natural versus anthropogenic forcing	179
	7.4 Energy balance models and glacier variations	180
	7.5 Global circulation models (GCMs)	180
	7.6 Future research priorities	182

References	**184**
Index	**201**

Preface

Although a number of books on environmental change refer to the glacial record (for example, Embleton, C. and King, A.M. (1975): *Glacial Geomorphology*, Edward Arnold; Bradley, R.S. (1985): *Quaternary Paleoclimatology*, Allen & Unwin; Drewry, D. (1986): *Glacial Geological Processes*, Edward Arnold; Oerlemans, J. (1989): *Glacier Fluctuations and Climate Change*, Kluwer Academic Publishers; Bradley, R.S. (1991): *Global Changes of the Past*, UCAR, Boulder, Colorado; Eddy, J.A. and Oeschger, H. (1991): *Global Changes in the Perspective of the Past*, John Wiley & Sons; Bell, M. and Walker, M.J.C. (1992): *Late Quaternary Environmental Change*, Longman Group; Dawson, A.G. (1992): *Ice Age Earth*, Routledge; Lowe, J.J. and Walker, M.J.C. (1992 and 1997): *Reconstructing Quaternary Environments*, Longman Scientific & Technical; Mannion, A.M. (1992): *Global Environmental Change*, Longman Scientific & Technical; Kemp, D.D. (1994): *Global Environmental Issues*, Routledge; Roberts, N. (1994): *The Changing Global Environment*, Blackwell; Houghton *et al.* (1995): *Climate Change 1995*, Cambridge University Press; Menzies, J. (1995): *Modern Glacial Environments*, Butterworth-Heinemann; Menzies, J. (1996): *Past Glacial Environments*, Butterworth-Heinemann; and Benn, D. I. and Evans, D.J.A. (1998): *Glaciers and Glaciation*, Edward Arnold), there is no book that concentrates on this specific theme. The relevant material is scattered in more general texts, especially on glacial processes and glacial stratigraphy. Thus, the book should appeal to a wide readership ranging from students to scientists in a number of disciplines related to climate and environmental issues.

The book is intended to be an authoritative work at an advanced level, accessible to higher level undergraduates, and an inspiration to young researchers interested in the subject of environmental change. Students taking advanced options in glaciology, glacial geomorphology and environmental change will hopefully find this book valuable. Such courses are mainly offered in Geography and Environmental Studies degree schemes, although university Geology/Earth Science departments are increasingly teaching environmental courses. The book is intended for use in physical geography, geological sciences, environmental sciences, glaciology, Quaternary science, and palaeoclimatology. The book will also be useful as part of the scientific basis for understanding recent environmental change.

Evidence derived from glaciers that may be used in palaeoclimatic reconstruction is widespread and increasingly diverse. A book that brings together these proxy data, explains their application and draws conclusions about the past, present and potential future climate change in relation to glaciers therefore fills a niche.

The main aim of this book has been to provide a thorough, up-to-date account of glaciers and ice sheets as monitors and indicators of environmental change. The book is intended to cover the record of environmental change within glaciers and ice sheets, and the record of past environments left by retreating glaciers. These themes are examined within the context of environmental change in general and global climate change in particular. Methods of palaeoenvironmental records are assessed and implications for future environmental change are discussed. Evidence from glacier ice, or left by glaciers in the landscape or within the geological record, provides one of the most important sources of information on environmental change.

Acknowledgements

We wish to express our gratitude to the Editors of this series – Ray Bradley, John A. Matthews, Neil Roberts and Martin Williams – and the Edward Arnold commissioning editor responsible for the Key Issues in Environmental Change series, Laura McKelvie, for inviting us to write this book on glaciers and environmental change. Our thanks also go to Luciana O'Flaherty for her patience during the completion of the book. Special thanks are due to John A. Matthews for numerous discussions both in the office and in the field about different aspects of Holocene glacier variations and environmental change. We also wish to thank Colin K. Ballantyne and Danny McCarroll for stimulating discussions during joint fieldwork both in Norway and Scotland. Our colleagues at the Geology and Geography departments at the University of Bergen have encouraged us during the writing process. Finally, we would like to thank our families for their patience during the work involved with this book.

1

Introduction

1.0 Chapter summary

Evidence from glacier ice, or left by glaciers in the landscape or within the geological record, provides one of the most important sources of information on environmental change. The Earth is a dynamic and constantly changing system, in which all components interact. Research on global environmental change aims to understand how these complex systems interact, and to identify linkages between them. This research may provide the basis for predicting future global environmental changes and their human consequences. Glacier monitoring using satellites, based on 20 years of space-based observations of glaciers, has been developed to build a database covering most glaciers of the world and to monitor changes in glaciers on a periodic basis.

1.1 The significance of environmental change

The Earth is a dynamic, non-static and constantly changing system, in which all components (the atmosphere, geosphere, cryosphere, hydrosphere, and biosphere including mankind) interact. Research on global environmental change tries to understand how these complex systems interact, and to identify the nature of linkages between them. This research may ultimately provide the basis for predicting future global environmental changes and their human consequences. To be able to

forecast future environmental changes, an understanding of past environmental changes is essential. Such research concentrates on the time evolution of processes operating on time scales from decades to millennia and their interactions through time. Hopefully this will enable assessments of causes and effects in this complicated dynamic system. Palaeoenvironmental studies also show how rapidly Earth systems may respond to forcing factors, which is important in planning future environmental change. This research provides a database of environmental conditions in the past which can be used for testing numerical models of atmospheric, terrestrial and marine processes.

For the time prior to instrumental records, evidence of environmental change comes from 'natural archives' or *proxy* records (Table 1.1), providing information about palaeoenvironmental conditions including past atmospheric composition, tropospheric aerosol loads, explosive volcanic eruptions, air and sea temperatures, wind and precipitation patterns, ocean chemistry and productivity, sea-level changes, ice-sheet dimensions, and variations in solar activity. Of crucial importance, however, is the ability to date the different records accurately in order to determine whether events occurred simultaneously, or whether events led or lagged behind others.

Glaciers and ice sheets are some of the best archives of past environmental change, as demonstrated by ice cores obtained from the Antarctic and Greenland ice sheets (see Chapter 3) and through the history of glacier

TABLE 1.1 Characteristics of natural archives (adapted from Bradley and Eddy, 1991)

Archive	Temporal resolution	Temporal range (yr)	Information
Historical records	Day/hour	1000	T H B V M L S
Tree rings	Season/year	10,000	T H C_a B V M S
Lake sediments	1 to 20 years	10,000–1,000,000	T H C_w B V M
Ice cores	1 year	100,000	T H C_a B V M S
Pollen	100 years	100,000	T H B
Loess	100 years	1,000,000	H B M
Ocean cores	100–1000 years	10,000,000	T C_w B M
Corals	1 year	100,000	C_w L
Palaeosols	100 years	100,000	T H C_s V
Geomorphic features	100 years	10,000,000	T H V L
Sedimentary rocks	1 year	10,000,000	H H C_s V M

T: temperature; H: humidity or precipitation; C: chemical composition of air (C_a), water (C_w), or soil (C_s); B: biomass and vegetation patterns; V: volcanic eruptions; M: geomagnetic field variations; L: sea-level; S: solar activity.

fluctuations obtained from the glaciated regions of the world (see Chapter 5). Records of glacier fluctuations contribute important information about the range of natural variability and rates of change with respect to energy fluxes at the Earth's surface over long time-scales. Reconstructed Holocene and historical glacier fluctuations indicate that the glacier extent in many mountain ranges has varied considerably during recent millennia and centuries, exemplified by the Little Ice Age and late-twentieth century weather extremes. The general shrinkage of Alpine glaciers during the twentieth century is a major reflection of rapid change in the energy balance at the Earth's surface. An annual loss of a few decimetres of glacier ice depth is largely consistent with the estimated anthropogenic greenhouse forcing (a few W/m^2). The rapid glacier retreat in the first half of the twentieth century was probably little affected by emissions of greenhouse gases. The later general retreat may, however, include an increasing component of human influence. Recent glacier shrinkage may now coincide with increased human-induced radiative forcing. Glacier mass balance measurements therefore become one of the key indicators for evaluating possible future trends.

1.2 Glaciers as monitors of environmental change

Glaciers and ice sheets are commonly located in remote areas far from population centres. Despite this physical and mental distance, past and modern glacial environments provide an important key to our knowledge of past, present and future global environmental conditions. Glacial environments may at first look chaotic and complex. However, few other environments exhibit such rapid, dynamic and spatially variable changes of processes. Past and present glaciers and ice sheets have had a significant impact upon all aspects of Earth systems. Understanding of many aspects of glaciers and glacier processes still remains poor. As an example, the complex relationship between ice dynamics and mass balance fluctuations is not fully understood. Modelling of ice masses and mass balance studies have, and will, advance our understanding of global ice-sheet fluctuations in the past.

The effect of modern glaciers on a global scale can be looked upon at two levels. Firstly, they impact upon humans and habitats in their nearby surroundings. Meltwater outbursts and rapid ice advances resulting in the loss of pasture lands, property and human fatalities are well documented (e.g. Grove, 1988). Secondly, there is the large-scale impact on the global climate and sea-level. Related to this topic is the controversial question of ice-sheet stability and whether the large ice sheets are melting at an increased rate, injecting large volumes of cold fresh water into the polar oceans, and affecting oceans and near-shore habitats, currents, surface ocean water temperatures and global weather phenomena.

Over the past several decades the techniques of studying glaciers have greatly improved. For example, satellite images have improved the accuracy of measuring ice movement and mass balance. Ice cores retrieved from the Antarctic and Greenland ice sheets have greatly improved our knowledge of past environmental changes. Computer-generated ice-sheet models have increased our understanding of ice-sheet growth and potential stability/instability as a result of predictions of future ice-sheet variations. In addition, there is growing knowledge of the likely spatial and temporal development of the Pre-Pleistocene and Pleistocene ice sheets, and the causative mechanisms that may lead to global glaciation, carbon dioxide variations, and biomass and productivity changes.

International monitoring of glacier variations began in 1894. At present, the World Glacier Monitoring Service (WGMS) of the International Commission on Snow and Ice (ICSI/IAHS) collects standardized glacier information, as a contribution to the Global Environment Monitoring System (GEMS) of the United Nations Environment Programme (UNEP) and to the International Hydrological Programme (IHP) of the United Nations Educational, Scientific and Cultural Organization (UNESCO). The database includes observations on changes in length and, since 1945, mass balance. Most of the data come from the Alps and Scandinavia.

Two main categories of data – summary information and extensive information – are reported in the glacier mass balance bulletins published by IAHS. Summary information on specific balance, cumulative specific balance, accumulation area ratio (AAR) and equilibrium

line altitude (ELA) is given for ca. or approximately 60 glaciers. This information provides a regional overview. In addition, extensive information such as balance maps, balance/altitude diagrams, relationships between accumulation area ratios, equilibrium line altitudes and balance, as well as a short explanatory text with a photograph, are presented for 11 selected glaciers with long and continuous glaciological measurements from different parts of the world. The long time series are based on high-density networks of stakes and firn pits. These data, most of which are now available on Internet, are useful for analysing processes of mass and energy exchange at the glacier–atmosphere interface and for interpreting climate/glacier relationships.

Glacier monitoring using satellites, based on 20 years of observations of glaciers by LANDSAT, SPOT, ERS and, in the future, EOS and Radarsat, has been developed to build a database covering most glaciers of the world and to monitor glacier changes on a periodic basis. Satellite monitoring of the world's glaciers should produce a uniform image data-set, monitor special events such as glacier surges, produce maps of the areal extent of glaciers and snow fields, give information about glacier surface velocities, advance and retreat, and an inventory of the glaciers, including mean surface speed, length, width, areal extent, snowline, and the temporal changes in these parameters. Adam *et al.* (1997) evaluated the effectiveness of ERS-1 synthetic aperture radar (SAR) imagery for mapping movement of the transient snowline in a temperate glacier basin during the ablation season. Despite localized confusion between glacier ice and wet snow, the wet snowline can be mapped reasonably well by using ERS-1 SAR imagery.

Studies show that most Arctic glaciers have experienced negative net surface mass balance over the last few decades (e.g. Dowdeswell *et al.*, 1997; Pohjola and Rogers, 1997a,b). There is, however, no uniform recent trend in mass balance in the Arctic, although some regional trends are recognizable. In northern Alaska, for example, glaciers experience increased negative mass balance as a result of higher summer temperatures. This development may

Box 1.1 The concept of ice ages – historical background

Environmental change is a continuous process where dynamic systems of energy and material operate on a global scale to cause gradual and sometimes catastrophic changes in the atmosphere, hydrosphere, lithosphere and biosphere. During most of the Earth's history the agents in the environmental system have been the natural elements (wind, ice, water, plants and animals). Some 2–3 million years ago, however, a new and perhaps the most powerful generator of environmental change, the hominids, emerged. The earliest testament to this are the cave paintings in many parts of the world. The first written accounts came with the rise of the ancient Mediterranean civilizations in Greece and Rome. Ideas changed little during the Middle Ages, when European scholars returned to the concept of a flat Earth. The bipartite nature of geography, first intimated in Strabo's work, involved human and physical divisions. This concept was formalized by Varenius (AD 1622–1650) who originated the ideas of regional or 'special' geography and systematic or 'general' geography. The deductive and mechanistic philosophy earlier advocated by Newton (1642–1727) was continued in the work of Charles Darwin (1808–1882). In his classic work *The Origin of Species* (1859) he advanced theories of evolution and suggested a relationship between environment and organisms. By the end of the nineteenth century, the theory of the biblical flood as a major agent in shaping the face of the Earth was questioned.

The earliest descriptions of glaciers are in Icelandic literature and date from the eleventh century. During the Little Ice Age, glaciers around the world expanded considerably. In the Alps and in Norway the glacier advance led to destruction of pastures and property. The ice-age theory was developed during the nineteenth century. The main spokesman for the theory of ice ages in the early nineteenth century was Louis Agassiz, the influential president of the Swiss Society of Natural Sciences, who has been regarded as the 'Father of Ice Ages'. Agassiz was, however, not the first to believe that glaciers had previously been more extensive, and he himself was sceptical for several years. Perhaps the first to document the evidence for more extensive glaciers was the Swiss minister Kuhn. In 1787 he interpreted erratic boulders below the glaciers near Grindelwald as evidence for a more extensive glaciation. Scot Hutton, one of the leading contemporary geologists, published in 1795 his 'Theory of the Earth' in which he described how ice had transported great boulders of granite into the Jura Mountains. A Swiss mountaineer and hunter named Perraudin argued in 1815 that glaciers had flowed into the Val de Bagnes in the Alps, and tried to convince Carpentier, who later became an advocate of the glacial theory, of his views. Perraudin also tried to persuade the Swiss engineer Venetz three years later, but he too was sceptical of the theory. However, Venetz began to accept the hypothesis and in 1829 he argued from the distribution of moraines and erratics that glaciers had covered the Swiss plain, the Jura and other regions of Europe. In 1824 the Norwegian geologist Esmark had already argued that glaciers in Norway had been much more extensive than at present. It was, however, the German poet Goethe who promoted the idea of an ice age (*Eiszeit*) in the novel *Wilhelm Meister* (1823).

Meanwhile, Carpentier accepted Venetz's theory of more extensive ice, and started to collect evidence in favour of this hypothesis. At that time it was believed that the biblical flood explained the distribution of the erratics, and resistance was therefore strong to the ice age concept. In 1833 several researchers had accepted the view of Lyell, the leading British geologist of the day, that boulders had been deposited by icebergs, a theory (the 'drift' theory) developed in 1804 by the German mathematician Wrede. Darwin supported Lyell, and in a series of papers he advocated the theory until his death in 1882.

During a field trip to Bex, Agassiz was convinced by Carpentier of the truth of the glacial theory, which for the first time had a strong, forceful and influential spokesman. By now Agassiz and Carpentier were familiar with Goethe's great ice age theory, but researchers ignored it because of his lack of scientific style. Unfortunately, Agassiz developed the glacial theory beyond available evidence, and when he presented the theory to the Swiss Society of Natural Sciences at Neuchâtel in 1837, he was met with great opposition. Agassiz published his work in 1840 in the book *Etudes sur les Glaciers*. The ice age theory substituted the Great Flood, of which Buckland, a professor of mineralogy and geology at Oxford University, was a great spokesperson. Buckland joined Agassiz on a trip to the Alps, but Buckland was still not convinced. After Buckland had discussed glacial deposits in Scotland and northern England with Agassiz, Buckland finally became convinced about the glacial theory. Agassiz moved to the US in 1847 as professor at Harvard University. Many researchers had already accepted his theory, but his appointment speeded up its acceptance, and when Agassiz died in 1873, only a few scientists had not yet accepted the ice age theory. Subsequently, tillites were found as evidence of ancient glaciations. Around the turn of the century, evidence for four ice ages were found in North America, the European Alps, Scandinavia, Britain and New Zealand. The first deep-sea sediment cores, covering most of the Quaternary, were

obtained in the 1950s. Oxygen isotope studies of planktonic and benthic foraminifera were used to estimate palaeotemperatures and ice volumes. In the early 1970s it was assumed that the period of glaciations was equivalent to the Quaternary period (ca. 2.5 million years). In 1972, long cores retrieved from the Antarctic continental shelf in the Ross Sea showed evidence of glaciations 25 million years ago. Cores obtained in 1986 showed evidence of glaciations as far back as 36 million years ago (Oligocene). The precise timing of the onset of Cenozoic glaciation in Antarctica remains to be determined.

As evidence of environmental change accumulated, attention also focused on the underlying cause of climate change. The French mathematician Adhémar was the first to involve astronomical theories in studies of the ice ages. In 1842 he proposed that orbital changes may have been responsible for climatic change of such magnitude. The Scottish geologist James Croll advanced a similar approach in 1864, suggesting that changes in the Earth's orbital eccentricity might cause ice ages. In the book *Climate and Time* he explained the theory in full. Due to the inability to date and test Croll's hypothesis, his theory was not seriously considered until Milutin Milankovitch, a Serbian astronomer, revived the theory during 1920–40. The Milankovitch theory, or the *astronomical theory* of ice ages, has become widely accepted since the 1950s with evidence from the deep-sea records. The late eighteenth and early nineteenth centuries witnessed the establishment of new methods, mainly based on biological remains, thus establishing the field of *palaeoecology*.

be a response to a step-like warming of the Arctic in the early twentieth century since the end of the Little Ice Age. Maritime Scandinavian and Icelandic glaciers, on the other hand, show increasingly positive mass balance due to increased precipitation during the accumulation season (Pohjola and Rogers, 1997a).

A degree-day glacier mass-balance model was applied by Johannesson *et al.* (1995) to three glaciers in Iceland, Norway and Greenland, where mass-balance data for several years are available. The model results corresponded reasonably well with measured variations in the mass balance with elevation for each glacier. A similar degree-day model approach was used by Braithwaite (1995) to study ablation on the Greenland ice sheet.

At the margin of the eastern North Greenland ice sheet, Konzelmann and Braithwaite (1995) studied variations in ablation, albedo and energy balance. Their results showed that net radiation is the main source for ablation energy, and turbulent fluxes are about three times smaller energy sources, while heat flux into the ice is a substantial heat sink which reduces the energy available for ice melt. Studies show that small-scale albedo variations must also be evaluated carefully in large-scale energy balance calculations.

1.3 The study of glaciers and past glacier fluctuations in the context of present and future environmental change

The past 2–3 million years (the Quaternary Period) have been characterized by periodic climatic variations. During cold periods, glaciers and ice sheets became more extensive than today, while in milder intervals, the glacier extent was much less. One of the main achievements of the earth sciences has been the demonstration that the sequence of *glacials* and *interglacials* are primarily driven by Earth's orbital parameters (Imbrie and Imbrie, 1979; Berger, 1988; Imbrie *et al.*, 1992, 1993a,b). This *external forcing mechanism* causes responses and chain reactions in the internal elements (atmosphere, oceans, the hydrological cycle, vegetation cover, glaciers and ice sheets) of the Earth (e.g. Bradley, 1985). Changes in one element of the Earth's system can cause responses in other elements because they are coupled in a linked system. These can lead to feedback reactions which can amplify the original signal. Glaciers and ice sheets play an important role in the global climate system. Glacier advance and retreat may therefore be both a consequence and a cause of climate change (Imbrie *et al.*, 1993a,b).

Ice sheets normally take longer to grow than to decay, but most information is available about their decay phases. This is mainly due to the fact that geomorphological and stratigraphical evidence of glacier retreat has a much better preservation potential than evidence from glacier build-up (e.g. Clark *et al.*, 1993).

Models of ice-sheet growth are largely theoretical because of little data. Flint (1971) proposed the highland origin, windward growth model for the Laurentide ice sheet in North America (Fig. 1.1). Alternative models suggest the coupling of regional-scale topography and climate. Ives *et al.* (1975) refined a model referred to as *instantaneous glacierization* for the growth of the Fennoscandian and Laurentide ice sheets. This model infers that snow accumulating on mountain plateaux produces plateau glaciers which may expand and coalesce to produce a multi-domed ice sheet.

The growth of ice sheets with large marine-based components was explained by Denton and Hughes (1981) in the *marine ice transgression hypothesis*. In this model, sea ice expands to inter-island channels and large embayments. The albedo increases, thus reducing the temperature and lowering the snowline. Snow accumulates to build ice shelves which eventually ground to form a marine ice dome.

There is a growing body of data from both terrestrial and marine environments that the large northern hemisphere ice sheets were characterized by relative instability during the last glacial cycle. Samples from the Hudson Bay area (e.g. Clark *et al.*, 1993) and sediment cores from the North Atlantic contain layers of lithic fragments ('Heinrich layers') interpreted as *ice-rafted debris* carried by icebergs (e.g. Heinrich, 1988). These ice-rafting events record episodes of ice break-up along the eastern margin of the Laurentide ice sheet, and indicate that even large ice sheets are able to respond rapidly to climatic and/or dynamic forcing and undergo large volume changes over a few thousand years.

When ice sheets are established, they influence regional climate and commonly create their own weather system. On minor glaciers and ice caps, precipitation normally increases with increasing altitude. On larger ice sheets, however, precipitation increases with altitude around the outer margins of the ice sheet.

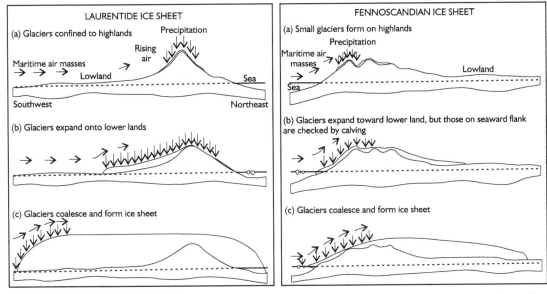

FIGURE 1.1 The model of highland origin, windward growth of ice-sheet inception. Adapted from Flint (1971), Ives *et al.* (1975) and Benn and Evans (1998).

Over the central parts, a high-pressure zone is the prevailing situation, which reduces precipitation considerably. In Antarctica, for example, the precipitation around the margins is about ten times that of the interior. High albedo and high elevation of the central parts of the ice sheets make the winter temperatures fall to around $-70°C$ in Antarctica and $-40°C$ over Greenland. Because an ice sheet reduces the absorption of solar radiation, the thermal contrasts between polar and equatorial regions are accentuated. Increased equatorial–pole temperature gradients during glaciations will increase the strength of the zonal (east–west) and meridional (north–south) circulation. High-latitude ice sheets can persist in a state of disequilibrium with existing climate because ablation rates are low. The West Antarctic ice sheet, which is grounded in the sea and has ice shelves in the Ross and Weddel seas, however, responds mainly to changes in glacial–interglacial sea-level fluctuations. The Antarctic ice sheet has remained relatively stable since the Miocene more then 20 Ma ago (Denton *et al.*, 1993). Global climate variations produced only minor changes at the margins of the Antarctic ice sheet compared with the oscillations of the Laurentide and the Eurasian ice sheets. The Greenland ice sheets have fluctuated considerably more than the Antarctic ice sheet, as demonstrated by the results from the GRIP and GISP2 ice cores (see Chapter 3).

Glaciers are sensitive to climate changes of various magnitudes and different time-scales. They therefore constitute an important source of palaeoclimatic data. In addition, their widespread geographical distribution makes them suitable for establishing climate proxy data and for evaluating the nature of global climate fluctuations (e.g. Porter, 1981a,b).

Because glaciers respond to changes in their climatic environment by growing or shrinking, they can be used as sensitive palaeoclimatic indicators (winter precipitation, summer temperature, and prevailing wind direction). Studies of ice cores give one of the best available indicators of climate variations back in time in polar and alpine regions (e.g. Johnsen *et al.*, 1992). Glacial deposits may form the basis for discontinuous time series of glacier fluctuations. However, glacier surges not related to climate, lags in the dynamic response of the glacier front to climatic variations, together with chronological uncertainties related to dating problems, make assessment of glacial geological data difficult.

The early to middle Holocene was, in general, a time of glacial retreat and warmer climate, termed the 'hypsithermal' (Deevey and Flint, 1957). The late Holocene witnessed the rebirth and readvance of most alpine glaciers throughout the world, an event termed Neoglaciation (Porter and Denton, 1967). During the Little Ice Age (last four to five centuries) there is widespread evidence for repeated glacier fluctuations throughout the world (e.g. Grove, 1988). During this time, the equilibrium line altitude (ELA) was lowered by approximately 100–200 m, equivalent to about 15 per cent of the ELA lowering at the last glacial maximum.

Historical observations going back to the eighteenth or nineteenth centuries help in reconstructing glacier fluctuations, together with terminal moraines dated by lichenometry and/or dendrochronology. Analyses of sediment cores from lakes downstream of glaciers may provide continuous records of Holocene glacier fluctuations. The relative abundance of minerogenic and organic content in lacustrine sediments is interpreted to be an indicator of glacier activity in the catchment area. Greater minerogenic content is generally indicative of more extensive glaciation (e.g. Karlén, 1976, 1981). Detrital organic matter, tephra and varves usually provide age control, and lake sediments are commonly in agreement with moraine records (e.g. Leonard, 1986).

Information about the Earth's climate in the past can help us to predict the direction and magnitude of future climatic change. This palaeoclimate information is inferred from a diverse array of biological, chemical and geological indicators (e.g. pollen, shells of marine micro-organisms, and glacial landforms).

2

Theories for climate and glacier variations

2.0 Chapter summary

The aim of this chapter is to demonstrate the contribution of glacier research to the development of climate change theories. Firstly, the astronomical (Milankovitch) theory of climate variation is explained. Spectral analysis of long marine and terrestrial climatic records have revealed cycles of 100,000, 41,000 and 23,000/19,000 years. Different data sets have confirmed the hypothesis that changes in the orbital variables comprise the primary forcing mechanism for Quaternary climatic changes. In addition, the relationship between variations in atmospheric gas content and climate change is assessed. Furthermore, the evidence of volcanic activity and its effect on climate variations is discussed. Finally, variations in solar output and geodynamic factors are discussed in relation to climate and glacier variations. It is concluded that variations in atmospheric gas content, volcanic aerosols and solar irradiation are significant contributors to climate change. However, the cyclic nature of the most significant climatic variations during the Quaternary is difficult to explain entirely by geodynamic factors.

2.1 The astronomical (Milankovitch) theory of climate variation

During at least the last million years, climate has fluctuated in a distinctive way. In recent years there have been attempts to explain the causes of the long-term climatic fluctuation (e.g. Imbrie and Imbrie, 1979; Bradley, 1985). The *astronomical theory* or the *Milankovitch theory* has undoubtedly attracted the greatest attention. This theory was in fact developed by Croll about 100 years ago, but was later elaborated by the Serbian geophysicist Milutin Milankovitch. The theory is based on the assumption that changes in the Earth's orbit and axis cause surface temperature changes on the Earth. Due to the gravitational influence of other planets, the shape of the Earth's orbit changes from almost circular to elliptical and back again over a period of approximately 100,000 years, a process referred to as the *eccentricity of the orbit*. (Fig. 2.1).

The tilt of the Earth's axis varies from $21°39'$ to $24°36'$ and back over a period of 41,000 years, called the *obliquity of the ecliptic*. The third factor, caused by gravitational pull exerted by the sun and the moon which causes the Earth to wobble around its axis, is termed *precession of the equinoxes* or *precession of the solstices*. This means that the seasons when the Earth is nearest to the sun (*perihelion*) varies with cycles of 23,000 and 19,000 years. At present, the northern hemispheric perihelion occurs in winter, while *aphelion* (Earth on the farthest point on the orbit) is in summer.

The total amount of radiation to the Earth is mainly determined by the eccentricity of the Earth's orbit. The other astronomical variables, however, affect the latitudinal energy distribution. The regularity of the astronomical effects makes it possible to calculate changes through

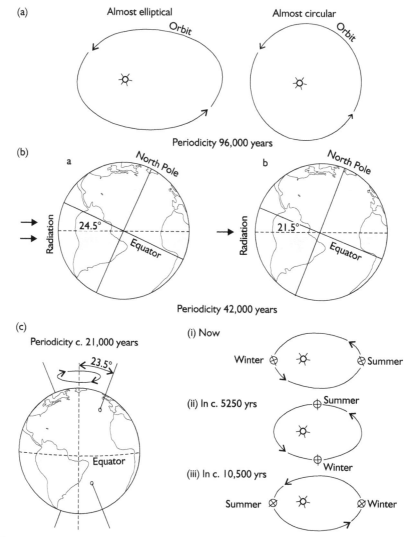

FIGURE 2.1 The components of the astronomical theory of climate change: (a) eccentricity of the orbit; (b) obliquity of the ecliptic; and (c) precession of the equinoxes. (Adapted from Imbrie and Imbrie, 1979).

time. Milankovitch was able to make radiation input estimates for different latitudes, and thereby temperature changes. Generally, solar radiation in the low and middle latitudes is mainly related to precession and eccentricity variations, while the effects of eccentricity are modified by obliquity changes in higher latitudes.

The astronomical theory was first published in 1924. Soon it became apparent, however, that the Late Quaternary glacial episodes, as reconstructed at that time, were not in accordance with the astronomical theory. In the mid-1950s the theory was more or less rejected. In the 1970s, however, studies of sea-level changes and deep ocean sediments led to increased interest in the Milankovitch theory. The $\delta^{18}O$ variations in marine microfossils, which record long-term environmental variations, made it possible to test the astronomical theory against climatic data. Spectral analysis of ocean climatic records (*oxygen isotope variations*) revealed cycles of 100,000, 41,000 and 23,000/19,000 years (Hays *et al.*, 1976). Evidence of the astronomical variables were subsequently found in various proxy records (coral reefs, pollen, loess, ice cores, lacustrine sediments). These data sets therefore confirmed the hypothesis that changes in the orbital variables are the primary forcing mechanism for Quaternary climate change (e.g. Imbrie *et al.*, 1993b).

Although the astronomical theory explains the main Quaternary climatic fluctuations, recent research has shown that other factors have also influenced global climate variation. Quaternary climatic cycles have not been constant. Prior to approximately 800,000 years ago, a periodicity of 41,000 years prevailed. Subsequently, the 100,000 yr climatic cycle dominated (Ruddiman *et al.*, 1986). During the last 700,000–800,000 years, northern hemispheric ice sheets grew larger than those attained during the previous 1.6–1.7 million years (Ruddiman and Raymo, 1988). Elements in the climatic system that may modify the orbital climate forcing include the location of landmasses, tectonic activity, oceanic circulation, ice cover, carbon dioxide, methane, and dust particles.

2.2 Variation in atmospheric gas content and climate change

Direct measurements of the atmospheric concentration of the greenhouse gas carbon dioxide (CO_2) began in 1958, and this figure has shown a significant increase from 315 parts per million by volume (ppmv) to 364 ppmv in 1997 (data from Keeling *et al.*, 1995; Keeling and Whorf, 1998; Fig. 2.2). The increase in atmospheric CO_2 is caused by burning of fossil fuels and a change in land use. Previous investigations showed that the atmospheric CO_2 concentration was about 280 ppmv

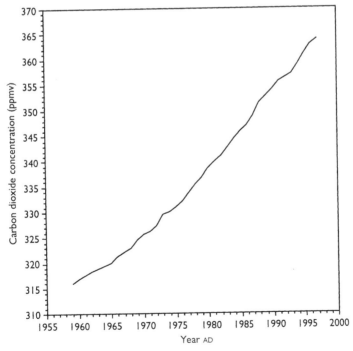

FIGURE 2.2 Atmospheric carbon dioxide concentrations (ppmv) derived from *in situ* air samples collected at Mauna Loa Observatory, Hawaii. (Data from C.D. Keeling and T.P. Whorf's (1998) web page)

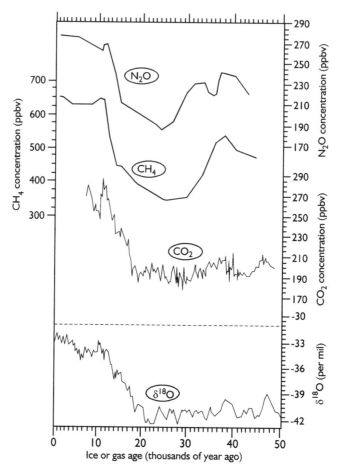

FIGURE 2.3 The Byrd ice-core record of CO_2, CH_4 and N_2O during the last 50,000 years, with the $\delta^{18}O$ climatic record for comparison. (Adapted from Raynaud *et al.*, 1992)

before industrialization, and that only minor changes of approximately 5 ppmv occurred throughout the pre-industrial part of the last millennium (Etheridge *et al.*, 1996). The atmospheric CO_2 concentration, however, increased from about 200 to 270 ppmv during the transition from the last glacial maximum (ca. 20,000 yr BP) to the beginning of the Holocene (11,000 yr BP) (Neftel *et al.*, 1988). Between 46,000 and 18,000 yr BP, fluctuations of about 20 ppmv occurred on a millennial time scale.

The causes of these variations are considered to reflect variations in production rates and operations of sources and sinks of different gases (Sundquist, 1993). During glacial periods of sparse vegetation cover, the oceans acted as a CO_2 sink. In contrast, warming of the oceans

in the Late-glacial and early Holocene caused CO_2 degassing to the atmosphere, where it was subsequently included in terrestrial vegetation and organic soils. Natural emission of methane occurs from peatland, animals, burning of organic material, oceans and lakes, of which variations in wetland extent and composition seem to have been the major source (Chappelaz *et al.*, 1990; Blunier *et al.*, 1995).

Records of gas concentrations from ice cores seem to correspond closely with temperature curves inferred from oxygen isotopes from the ice sheets (Fig. 2.3). Both carbon dioxide and methane are closely in phase with the climatic signal during deglaciation periods. At the beginning of a glacial, in contrast, methane is in phase but carbon dioxide lags behind (Raynaud *et al.*,

1992). Spectral analysis of the CO_2 and methane contents in the Vostok ice core (Antarctica) shows a maximum at ca. 21,000 yr BP and a weaker maximum at 41,000 yr BP (precessional and obliquity periodicities) (Barnola *et al.*, 1987; Lorius *et al.*, 1988; 1990). Variations in the CO_2 and methane concentrations are estimated to have caused about half of the temperature fluctuations recorded in the Vostok ice core (Lorius *et al.*, 1988, 1990). Fluctuations in the greenhouse gases probably represent a complicated response to orbitally driven climate changes, and are therefore important elements in the cause(s) of long- and short-term climatic variations.

2.3 Volcanic activity and climate variations

Explosive volcanic events inject fine-grained ash and dust into the atmosphere. This commonly leads to short-lived temperature drops due to reduction of incoming radiation. Dust particles may also act as foci for formation of water droplets and thereby cloud formation. Sulphur volatiles are, however, more significant in terms of climate change. In the atmosphere these are converted into sulphuric acid and these aerosols result in cooling of the lower troposphere by back-scattering of long wavelength radiation. The residence time of these aerosols is commonly between one and five years. After the Mt. Pinatubo eruption in the Philippines in 1991, solar radiation declined by up to 10 per cent and surface temperatures in the northern hemisphere dropped approximately 1°C (Handler and Andsager, 1994). Porter (1986) found temperature reductions following major volcanic eruptions of about 1.5°C during historical times.

Instrumental temperature records over the past 200 years and satellite data for the last 20 years have combined to show that explosive volcanic eruptions that emit large amounts of sulphur-rich gases into the stratosphere can reduce global temperatures by about 0.3°C over a period of 3–4 years following the eruption. The global cooling is not, however, homogeneously distributed. In addition, the number and type of eruptions that have occurred during the last 200 years is limited, and evaluating the impact of particular types of eruptions under different climatic modes is restricted to modern climatic conditions. Consequently, a complete understanding of the volcanism–climate system requires a multidisciplinary approach beyond the instrumental temperature data time period (e.g. Zielinski, 1998).

The volcanic records are developed through the evaluation of the direct products of the eruption, from terrestrial archives, ice core records, and atmospheric phenomena linked to the presence of aerosols in the stratosphere (e.g. red sunsets, dimmed lunar eclipses, dry fog). Volcanic records may in addition be deduced from proxy data (e.g. tree rings and coral records) of the climatic cooling resulting from a specific eruption. One of the most reliable records of past volcanic activity comes from continuous, high-resolution records of geochemical and conductivity variations in ice cores (see Chapter 3). Additional verification of source eruptions can be achieved through tephrochronological investigations. The stratospheric loading and optical depth for a particular eruption provide information for models and postulate the climate forcing of the eruption. High-resolution ice-core records from Greenland and Antarctica from the last 400 to 2000 years demonstrate the potential of eruptions to influence past climate and hence to modify future climate. In addition, these records show that several closely spaced eruptions may have a climatic impact on decadal time-scales. The ice-core records of volcanism also support the hypothesis that rapid climatic changes during glacial growth and decay periods can enhance crustal stresses, leading to increased periods of volcanic activity (Sejrup *et al.*, 1989).

Studies from terrestrial archives provide information on the composition of magma, volume erupted, and dispersal direction(s). Correlation of particular volcanic eruptions between terrestrial deposits, marine sediment cores, and ice cores provides distinct time markers for different proxy records. Compilations of volcanic records suggest that the global record of volcanism is incomplete, preventing prediction of the global and regional climatic impact of future volcanic eruptions.

A network of circum-hemisphere tree-ring density chronologies related to annual summer temperature variations has been combined into a single time-series for northern high latitudes and the northern hemisphere. Based on this well-dated, high-resolution composite time series, Briffa *et al.* (1998) suggested that large explosive volcanic eruptions produced cooling events in the northern hemisphere during the last 600 years. The significant temperature effect of some events, such as in 1816, 1884 and 1912, are apparent. The most severe short-term northern hemisphere cooling event of the past 600 years occurred in 1601, probably as a result of the AD 1600 eruption of the Huaynaputina volcano in Peru (Shanaka and Zielinski, 1998).

2.4 Variations in solar output

Since ancient times it has been known that the sun is the main source of energy driving the Earth's climate system (e.g. Gilliland, 1989). Changes in the solar constant were predicted by Milankovitch, who calculated the gravitational effect of the planets on the orbital parameters of the Earth. Until recently, it has been generally believed that the energy output from the sun is fairly stable. Since 1978, however, high-precision measurements of solar irradiance by satellite-based radiometers have shown that the solar constant has changed in phase with solar activity by 1–1.5 per mil. These measurements and observations raise three fundamental questions:

(1) How large can solar variability be on longer time scales?
(2) What is the sensitivity of the climate system to changes in solar forcing?
(3) To what extent are present and past climate changes caused by solar forcing?

Concerning question 1, observations of 20 solar type stars, exhibiting clear activity variations, indicate that changes of the solar energy output of about 1 per cent cannot be excluded, especially when considering longer time scales. The second question is difficult to answer, because the relative changes in the solar constant are significantly larger for shorter wavelengths. In addition, several processes within the atmosphere directly or indirectly affect the radiation balance and may amplify the solar effect. To the third question, there is growing evidence that periods with special climatic conditions have coincided with time intervals of extreme solar activity. Quiet sun periods, like the Maunder minimum, have tended to be colder and synchronous with the expansion of Earth's glaciers. Cosmogenic radionuclides (^{10}Be and ^{14}C) make it possible to extend the record of solar activity from a few centuries to several millennia, and to investigate solar and natural climate variations during recent millennia.

Variations in solar radiation have been regarded as a significant contributor to climate change. *Sunspots*, dark areas on the sun's surface, are indicators of changes in solar activity. Sunspot observations over the last two centuries indicate an 11 yr periodicity, with a longer but less pronounced 78 yr periodicity. Records of past solar changes include measurements of the cosmogenic isotopes of ^{14}C and ^{10}Be. ^{14}C production can be obtained by comparing radiocarbon dates from tree rings with the calendar ages (Stuiver *et al.*, 1991), while ^{10}Be production can be measured in ice cores (Beer *et al.*, 1992). Dansgaard and Oeschger (1989) found that the close correspondence between the two records may be attributed to solar output variations. Spectral analysis of the ^{14}C record suggests periodicities of 11 and 22 years (Hale cycles), an 88-year cycle (Geisberg cycle), and 200 and 2500-year cycles (Rind and Overpeck, 1993). From studies of ice cores (Dansgaard *et al.*, 1984), ocean cores (Pestieux *et al.*, 1987), tree rings (Sonett and Finney, 1990), varved sediments (Anderson, 1992) and lake-level variations (Magny, 1993), there seems to be some empirical support for a relationship between solar output and climate change. The *Maunder minimum*, the most recent episode of reduced sunspot activity, occurred during the coldest part of the Little Ice Age. During the Maunder minimum the reduction of solar insolation was in the order of 0.25 per cent, equivalent to a global temperature decline of around 0.5°C. Empirical evidence suggests, however, a temperature decline during the Little Ice Age in the range of 0.5–1.5°C (Rind and Overpeck, 1993). This suggests that other factors must be invoked to explain the Little Ice Age cooling, such as the North Atlantic thermohaline circulation (Stuiver and Brazunias, 1993).

Irradiance varies with sunspot number, but their direct climatic effect is minor over an 11-year sunspot cycle. Over the last 1000 years there have been periods when sunspot numbers were near zero: the *Wolf* (AD 1280–1350), *Spörer* (AD 1416–1534) and *Maunder* (AD 1654–1714) minima. Wigley (1988) related ^{14}C anomalies to glacier fluctuations throughout most of the Holocene. Wigley and Kelly (1990) found a statistically significant correlation between the global glacier advances of Röthlisberger (1986) and ^{14}C concentration during the Holocene. They interpreted the results, however, only as a strong indication since many uncertainties were involved in the analyses. Karlén and Kuyilenstierna (1996) compared Holocene climate changes in Scandinavia with changes in solar irradiation. For most of the last 9000 years they found a fairly good correspondence between cold events and δ^{14}C anomalies. The general Holocene cooling trend was suggested to be a combined result of land uplift after deglaciation and orbitally forced irradiation changes. Kelly and Wigley (1990) found that the influence of the enhanced greenhouse effect on global mean temperature dominated over the direct influence of solar variability. Prior to the period of direct sunspot observations, the differences between radiocarbon years and calendar years, as measured in tree rings, reflect anomalies in atmospheric ^{14}C concentration (e.g. Stuiver and Brazunias, 1993; Kromer and Becker, 1993).

Variations in solar radiation caused by changes in the Earth's orbital parameters (mainly precession) had significant effects on Holocene global climate (COHMAP Members, 1988). In the early Holocene this involved a greater input of solar radiation to the top of the atmosphere during northern hemisphere summers and southern hemisphere winters.

Maximum summer solar radiation at high latitudes of the northern hemisphere occurred at about 11,000 BP (Berger, 1978), giving summer solstice radiation 7–8 per cent greater than at present at 60°N. During the Holocene, summer insolation gradually reduced to present values. Prior to approximately 4500 BP, maximum insolation at 60°N occurred during mid-summer (June–July). Between 4500 and 4000 BP, however, maximum insolation at 60°N changed to late summer/autumn (August–September) (Berger, 1978).

A high-resolution oxygen isotope record (mainly reflecting temperature variations) of the GISP2 ice core from Summit Greenland was analysed for solar influences (Stuiver *et al.*, 1997). The atmospheric ^{14}C record was used as a proxy of solar change and compared with the oxygen isotope signal obtained from centimetre-scale isotope measurements from the period subsequent to AD 818. The analysis suggested a solar component to the forcing of Greenland climate during this millennium. The climatic response of the cold interval associated with the Maunder sunspot minimum, the Medieval warm period and the Little Ice Age temperature decline seem to have been related to solar climate forcing. For the rest of the Holocene, the oxygen isotope record shows more frequent fluctuations than the ^{14}C record. Ocean–atmospheric circulation forcing of climate may therefore have dominated over other forcing mechanisms during this interval.

2.5 Geodynamic factors

It has been postulated that changes in the distribution and volume of land ice and the rise and fall in eustatic sea-level during glacial–interglacial cycles could have affected the momentum of the Earth (deceleration of the Earth's rotation during periods of sea-level rise, and acceleration during low sea-level stands; Mörner (1993)). This could have led to variations in the direction and velocity of the major ocean currents, e.g. the North Atlantic Current (and the Gulf Stream), the Labrador Current and the Humbolt Current. Mörner

(1993) argued that such changes may explain some short-term global variations during the last millennium and at the Weichselian–Holocene boundary. However, due to lack of proxy data directly linked to geodynamic variations, it is difficult to test this hypothesis. In addition, the cyclic nature of the most significant climatic variations during the Quaternary is difficult to explain entirely by geodynamic factors. Changes in the spin velocity of the Earth and related effects may have contributed to modulating Late Quaternary climate changes (e.g. Lowe and Walker, 1997).

3

Methods of palaeoenvironmental reconstruction

3.0 Chapter summary

Long ice cores from Greenland and Antarctica and shorter cores from minor ice caps and glaciers have documented annual and decadal climate change during the last interglacial/ glacial cycle. The recent central Greenland ice cores (GRIP and GISP2) have clearly demonstrated the occurrence of large, rapid, regional to global climate oscillations during most of the last 110,000 years on a scale not recorded in modern times. Most of the glacial–interglacial changes occur over decades, while some indicators of atmospheric circulation change in only a few years. These millennial-scale events over Greenland were significant, with temperature fluctuations of up to 20°C, doubling of the snow accumulation, significant changes in wind-blown dust and sea-salt loading, and approximately 100 ppbv variations in methane (CH_4) concentrations. The recent central Greenland ice cores have given information about the origin of the ice sheet and its basal conditions, reconstruction of atmospheric circulation patterns and their temporal variations from chemical indicators and dust sources, and the anthropogenic influence on the chemical composition of the atmosphere. In addition, the cores have given data on glacier physics and flow modelling, solar influences on climate, and former size and atmospheric response of volcanic eruptions. In addition, the ice core from the Vostok station

in East Antarctica has allowed the extension of the ice record of atmospheric composition and climate to the past four glacial–interglacial cycles.

This chapter also reviews how glacier-front variations are mapped and measured, and dating techniques used on terminal moraines and other deposits in glacier forelands, such as observation and measurements, historical documents, biological dating methods and physico-chemical techniques, are assessed. The use of lacustrine sediments and laminae/ varves to extract the (palaeo)climate signal is also discussed. Finally, the great potential of marine sediments in climate reconstruction is described.

3.1 Stable isotope variations in ice cores

The atoms of oxygen (O) and hydrogen (H) present in glacier ice occur in different isotopes. Isotopes are different forms of an element that result from variations in atomic mass, or the combined number of protons and neutrons in each atomic nucleus. The number of protons in atoms of each element is constant. Mass variations are therefore a result of variations in the number of neutrons in the atom. Oxygen atoms have eight protons, but may have eight, nine or ten neutrons, giving three different isotopes with atomic

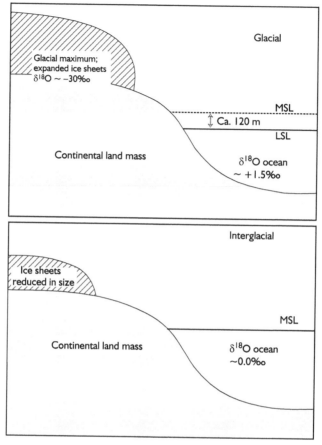

FIGURE 3.1 Variations in surface water oxygen isotope ratios during glacial maxima (low sea-level) and interglacials (maximum sea-level). (Adapted from Lowe and Walker, 1997)

masses of 16 (^{16}O), 17 (^{17}O) and 18 (^{18}O). Hydrogen atoms have one proton, but may have no or one neutron, giving two isotopes (^{1}H and ^{2}H), the latter also known as deuterium (D; 0.016 per cent). Water molecules may therefore consist of any of nine possible combinations of these five isotopes. Three of these combinations are, however, common: ^{1}H$_2$^{16}O, ^{1}HD^{16}O, and ^{1}HD^{18}O. The relative amount of oxygen isotopes in nature is 99.76 per cent ^{18}O, 0.04 per cent ^{17}O and 0.2 per cent ^{16}O.

The isotopic composition of precipitation falling on a glacier or ice sheet depends on the history of evaporation and condensation in the hydrological cycle. During the evaporation process, water molecules consisting of light isotopes turn to vapour more easily than those composed of heavy isotopes. This process is called *fractionation*. The resulting vapour is relatively enriched in ^{1}H and ^{16}O. As condensation proceeds, more of the remaining heavy isotopes will be removed, the vapour becoming more and more depleted in ^{18}O and D. Cooling of water vapour as it rises in the atmosphere and/or is transported inland over ice sheets will result in precipitation with increasingly lighter isotopic composition. As a result, the isotopic composition of the precipitation reflects the temperature when the precipitation occurred (Fig. 3.1). Measurements have demonstrated that there is a high correlation between temperature and oxygen isotope composition (Fig. 3.2), with a calibration of 0.33 per mil °C^{-1} (Cuffey *et al.*, 1995). Despite the fact that temperature is not the only factor determining the oxygen isotope

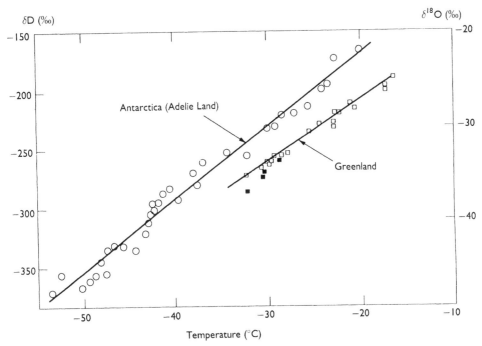

δD (‰)

$\delta^{18}O$ (‰)

FIGURE 3.2 Isotopic composition of snow versus local annual mean surface temperature. (Modified from Jouzel *et al.*, 1997)

composition, the isotopic composition of glacier ice has been used for temperature reconstructions. Jouzel *et al.* (1997) reviewed the empirical temporal slopes (curve gradients) in isotope models from polar regions. They found that the temporal slopes were lower than modern slopes, the difference most probably due to changes in the evaporative origins of moisture, changes in the seasonality of the precipitation, changes in the strength of the inversion layer, or a combination of these factors. Despite problems with calibrating an isotope palaeothermometer, the use of isotopes as a temperature proxy seems justified.

The oxygen isotope composition in ice cores, measured by mass spectrometry, is given as deviations ($\delta^{18}O$) from the *Standard Mean Ocean Water* (SMOW). Atmospheric water becomes depleted (average about 10 per cent) in the heavier ^{18}O isotope during evaporation from the ocean surface. There is, however, a seasonal variation of about 15‰ in the isotopic composition. Seasonal variations can therefore be detected by precise oxygen

isotope measurements. Due to diffusion effects, the amplitude of the $\delta^{18}O$ signal decreases with depth, but significant variations can still be detected back to 160,000 yr BP.

Hydrogen isotopes act in the same manner as oxygen isotopes. The hydrogen/deuterium ratio in the atmospheric water (snow) is determined by saturation vapour pressure and molecular diffusity in air. A deuterium profile from the Vostok ice core shows similar variations as in the Greenland oxygen isotope records (Jouzel *et al.*, 1990).

Downcore variations in stable isotope content are used to reconstruct the pattern and amplitude of climate variations. Figure 3.3 shows the oxygen isotope record from the GRIP core in Greenland and a deuterium profile from the Vostok core in Antarctica. Since isotopic fractionation is temperature dependent, the fluctuations mainly reflect global temperature variations. The curves are similar to those obtained from deep ocean cores. Both the Eemian and the Holocene are easily detected in the records.

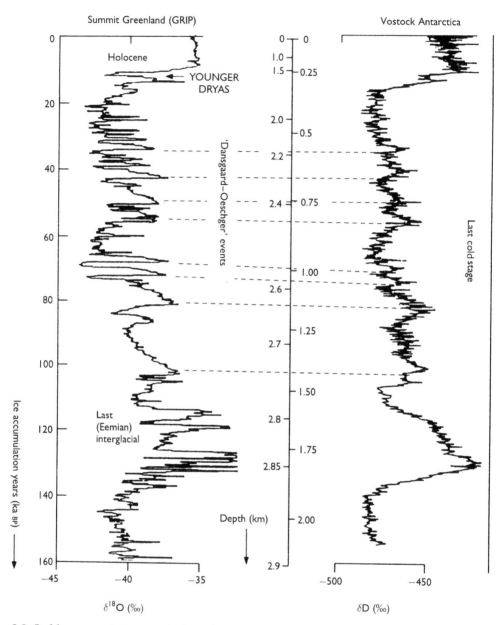

FIGURE 3.3 Stable oxygen isotope variations during the last 160,000 years recorded in the GRIP ice core, and deuterium ratios in the Vostok core from Antarctica. (Adapted from Lowe and Walker, 1997)

In the GRIP and GISP2 data records there are numerous, high-frequency $\delta^{18}O$ oscillations postdating the Eemian interglacial. Between 80,000 and 20,000 yr BP, some 20 interstadial events are recorded. These are interpreted to reflect abrupt temperature changes of the order of 5–8°C. These so-called *Dansgaard–Oeschger events* lasted for about 500–2000 years and therefore cannot be explained by orbital forcing mechanisms. Instead, they are interpreted to reflect feedback mechanisms involving ice sheet/glacier fluctuations, variations in the ocean system and atmospheric circulation fluctuations.

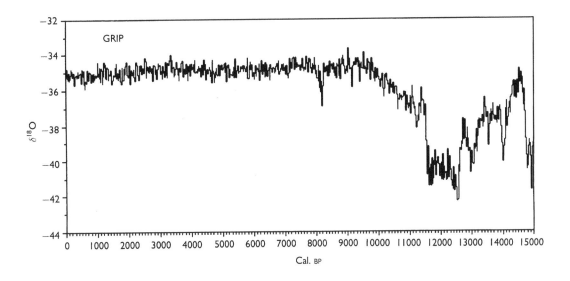

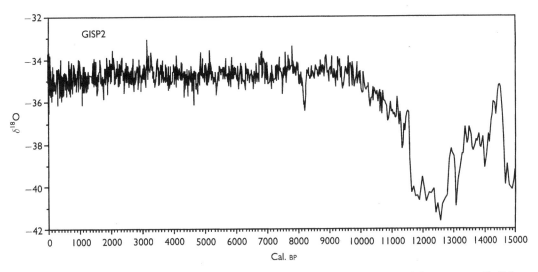

FIGURE 3.4 Late-glacial and Holocene oxygen isotope variations in the GRIP (upper panel) (Johnsen *et al.*, 1997) and GISP2 (lower panel) ice cores (Stuiver *et al.*, 1995).

Obtaining temperature series from stable isotope records from ice cores is not straightforward. This is especially the case for older parts of the record, because of glacier deformation and flow of ice from other regions. Different modern isotope values between the source area and the core site could give isotopic variations unrelated to real temperature variations. The GRIP and GISP2 cores were drilled close to the summit of the Greenland ice sheet, and if the ice divide has not moved significantly during the last interglacial/glacial cycle, ice flow has been minimal.

The relationship between oxygen isotopic ratios in precipitation and climatic conditions is also difficult to quantify (Lorius *et al.*, 1989). Several approaches have been proposed, including comparison of isotopic values in seasonal snow/ice with meteorological data (Dansgaard *et al.*, 1975), use of statistical methods to correlate borehole temperature records and isotope data, and constructing calibration models for oxygen isotope ratios to palaeotemperature (Cuffey *et al.*, 1992).

Interpretation of the isotopic signal in the GISP2 ice core indicates that the site is influenced by both the Icelandic Low to the SE and the Davis Strait/Baffin Bay storms to the SW and W (Barlow *et al.*, 1997). The GISP2 isotope signal is influenced by the North Atlantic Oscillation: the seesaw in winter temperatures between west Greenland and northern Europe.

Oxygen isotope profiles from the GISP2 summit area show rapid smoothing of the $^{18}O/^{16}O$ signal near the surface. Below a depth of about 2 m the smoothed $\delta^{18}O$ signal is fairly well preserved, interpreted to reflect average local weather conditions; the longer climate variations also have regional and global significance (Grootes and Stuiver, 1997). Between approximately 75,000 and 11,650 yr BP (the Younger Dryas/Preboreal transition) the oxygen isotope record is characterized by frequent, rapid switches between intermediate interstadial and low stadial values. Spectral analysis of the variations superimposed on the orbitally induced changes yields significant periodicities of 1500 and 4000 years. Similar fluctuations as recorded in the oxygen isotope signal in the GISP2 ice core have also been found in other climate records, strongly suggesting that the GISP2 oxygen isotope signal is the local expression of more regional and worldwide climate events. Meltwater from ice sheets adjacent to the North Atlantic influenced ocean circulation during the Bølling–Allerød–Younger Dryas complex of interstadials and stadials. The Holocene is characterized by relatively stable mean isotopic values (Fig. 3.4), however, with dominant 6.3, 11 and 210 year oscillations. The latter two are also recognized in the solar-modulated records of the cosmogenic isotopes ^{10}Be and ^{14}C, indicating that variations in solar irradiance is the main cause of these periodicities. Cooling by volcanic eruptions is recorded in the oxygen isotope signal; the effects, however, are small and volcanic eruptions are considered not to trigger large climate variations (Grootes and Stuiver, 1997).

The INTIMATE (INTegragion of Ice-core, MArine and TErrestrial records) group proposed that the GRIP ice core in Greenland be designated a stratotype for the Last Termination period (ca. 22,000–11,500 yr BP), and that the oxygen isotope profile be used as the basis for an event stratigraphy (Fig. 3.5) divided into stadials and interstadials according to their isotopic variations (Björck *et al.*, 1998).

3.2 Climate records from ice cores

Palaeoclimatic records can be obtained from ice cores drilled from glaciers and ice sheets, containing information on accumulation and atmospheric composition through time. In the upper part of ice cores, annual layers are commonly preserved as alternating bands of clear and bubbly ice. Deeper in the ice cores, however, annual layers are usually not discernible and dating is made indirectly. Short ice cores can be drilled manually, but long cores have to be retrieved by sophisticated mechanical equipment.

Annual layers in ice sheets and ice caps form as a response to winter accumulation and summer ablation. Near the glacier surface the winter layers are normally light in colour, while summer layers are darker due to partial melting and impurities. Deeper in the ice mass, however, the annual layers are difficult to detect due to thinning and distortion through pressure from the ice above and flow deformation. Analyses of ice cores from high altitudes and high latitudes, where little or no surface melting takes place during the ablation season, have demonstrated that these contain a great deal of palaeoenvironmental information. For example, annual layer thickness provides information about winter snowfall and degree of melting (determined by summer temperature). Aerosol and dust particles give information about the history of volcanic activity and dust storms in desert regions. Trace gases in the atmosphere are recorded in the composition of carbon dioxide and methane trapped in the englacial air bubbles. Stable isotopes (mainly oxygen isotopes) are proxies for climate change and act as correlation tools between terrestrial and

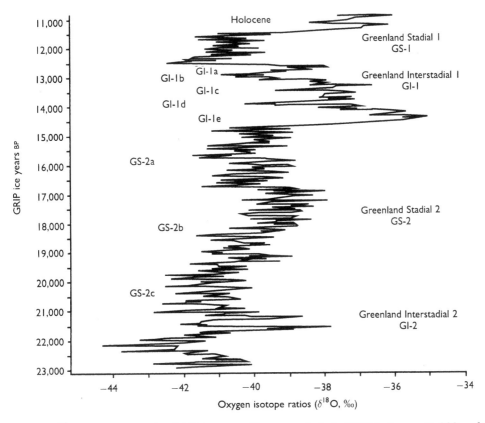

FIGURE 3.5 The $\delta^{18}O$ record from the GRIP ice core (Dansgaard *et al.*, 1993) between 11,000 and 23,000 years BP, with proposed stadials and interstadials. (From Björck *et al.*, 1998).

marine records. Natural and artificial radio-active isotopes provide the possibility of dating ice cores. The wide range of data obtained from ice cores therefore makes them one of the most important archives of Late Cenozoic palaeoenvironmental data.

Deep cores from Greenland (Fig. 3.6) and Antarctica and shorter cores from minor ice caps and glaciers have documented annual and decadal climate change beyond the last interglacial. The first ice cores were obtained from Camp Century (NW Greenland) in 1966, Dye 3 (south Greenland) in 1981, Renland (east Greenland) in 1988, and from Devon Island (North West Territories) in 1976. On the Antarctic ice sheet, cores were retrieved from Byrd Station (1968), Dome C (1979) and Vostok Station (1985). In the early 1990s, two cores were obtained from the summit of the

Greenland ice sheet. The *Greenland Ice Core Project* (GRIP) reached bedrock at 3029 m in 1992, while the North American *Greenland Ice Sheet Project Two* (GISP2) drilled c. 30 km away from the GRIP location, and reached bedrock at a depth of 3053 m in 1993. The GISP2 and GRIP ice cores, going back approximately 250,000 years, unequivocally demonstrated the presence of rapid climate-change events.

Initial interpretations of the GRIP ice core indicated that the rapid climate shifts present during the last ice age also persisted through the previous interglacial (the Eemian, Sangamonian, or isotope stage 5e). The GISP2 ice core also showed significant oscillations through the same period. The timing and character of the fluctuations were, however, different. Detailed analyses showed large structural disturbances caused by ice flow at

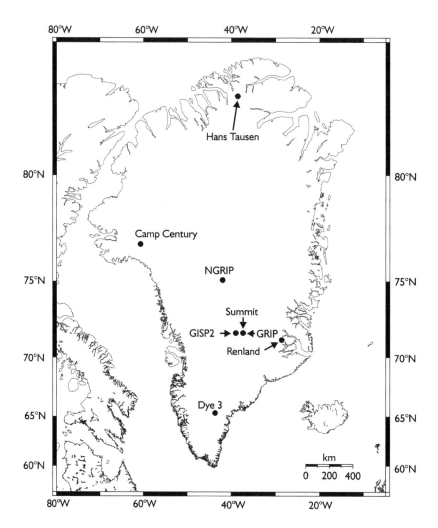

FIGURE 3.6 Six deep drilling sites in Greenland: Camp Century (US Army Cold Regions Research and Engineering Laboratory, 1966); Dye 3 (GISP, 1981); Renland (Nordic Council of Ministers, 1988); Summit (GRIP, 1992); the USA camp (GISP2, 1993); Hans Tausen (Nordic Environmental Research Programme, 1995); and the NGRIP started in 1996. (Adapted from Johnsen *et al.*, 1997)

about 2800 m depth in the cores, corresponding to an age of ca. 110,000 years (Alley *et al.*, 1997a). In addition, comparison with the undisturbed Eemian sequence in the Vostok ice core from Antarctica indicated that the sequence of ice layers older than 110,000 years are disturbed in both the GISP2 and GRIP ice cores (Chappelaz *et al.*, 1997).

In the GISP2 and GRIP cores it was possible to count annual layers into the glacial period (>10,000 years ago) and possibly down to

110,000 years, or for about 90 per cent of the core lengths (Fig. 3.7). (For visual-stratigraphic dating of the GISP2 ice core, see Alley *et al.*, 1997b.)

The age differences between dates of the cores and independent age markers are about 1 per cent in the Holocene and 5–20 per cent through most of the glacial period (Meese *et al.*, 1997). The GISP2 and GRIP cores show an almost perfect match back to 110,000 years ago. Volcanic markers and atmospheric

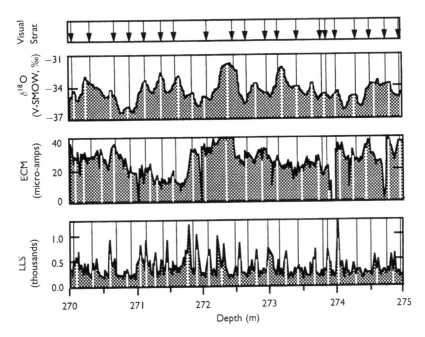

FIGURE 3.7 Multiparameter sequence between 270 and 275 m in the GISP2 ice core. Twenty stratigraphic layers and $\delta^{18}O$ peaks were counted in conjunction with 22 electrical conductivity method (ECM) and 23 laser-light scattering from dust (LLS) peaks. Annual layer markers are shown as vertical white lines, corresponding to the spring/summer inputs for each parameter. (Modified from Meese *et al.*, 1997)

oxygen ratios allowed correlation between the two ice-core records and with other ice core and marine deep-sea records. The central Greenland ice cores have clearly demonstrated the occurrence of large, rapid, regional to global-scale climate oscillations during most of the last 110,000 years, on a scale not recorded in modern times (Figs 3.8 and 3.9).

Most of the glacial–interglacial differences normally occur over decades, while some indicators of atmospheric circulation change in only 1–3 years. These millennial-scale events over Greenland were quite large, with temperature fluctuations of up to 20°C, doubling of the snow accumulation, significant changes in wind-blown dust and sea-salt loading, and approximately 100 ppbv variations in methane (CH_4) concentration. In addition, the analyses of the recent central Greenland ice cores have given information about the origin of the ice sheet and its basal conditions, reconstruction of atmospheric circulation patterns and their

temporal variations from chemical indicators and dust sources, and the anthropogenic influence on the chemical composition of the atmosphere. In addition, the cores have given data on glacier physics and flow modelling, solar influences on climate, and former size and atmospheric response of volcanic eruptions.

A study of the isotope and gas composition of the basal silty ice in the GRIP ice core indicates that local ice formed in the absence of the ice sheet is still preserved (Souchez, 1997). The ice probably formed in a peat deposit under permafrost conditions. This local ice was subsequently mixed with ice from an advancing ice sheet, according to the 'highland origin and windward growth' hypothesis for development of ice sheets (see Fig. 1.1).

In the GRIP core, continuous profiles of electrical conductivity were obtained (Fig. 3.10) (Wolff *et al.*, 1997). After having been corrected for temperature and density, the electrical conductivity reflects acidity variations, while

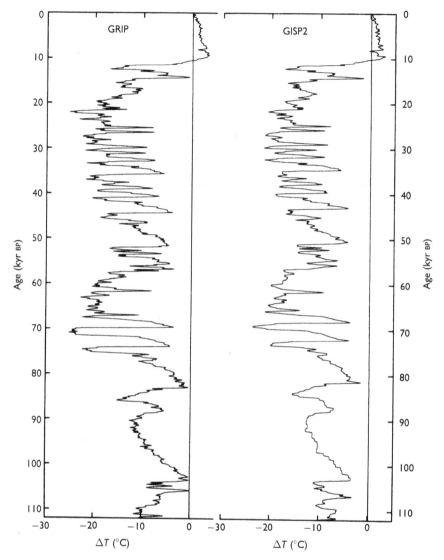

FIGURE 3.8 Calculated temperature change at the GRIP and GISP2 sites over the last 110,000 years. The GRIP data are from Johnsen *et al.* (1995), while the GISP2 data are from Cuffey *et al.* (1995). (Adapted from Jouzel *et al.*, 1997)

dielectric profiling yields acid, ammonium and chloride. Acidity dominates the variations in dielectric profiling during the Holocene, Allerød/Bølling, and larger interstadials. Ammonium dominates during the Younger Dryas, whereas chloride contributes most in cold periods and minor interstadials. The ice varies from acidic during the Holocene to

alkaline in the cold periods. During the interstadials, however, the ice is close to neutral.

Electrical conductivity measurements (ECM) in the GISP2 ice core reflect the ^+H concentration in the core (Taylor *et al.*, 1997). Seasonal variations in the nitrate concentration were used together with annual layer counting to date the core, and to correlate the GRIP and

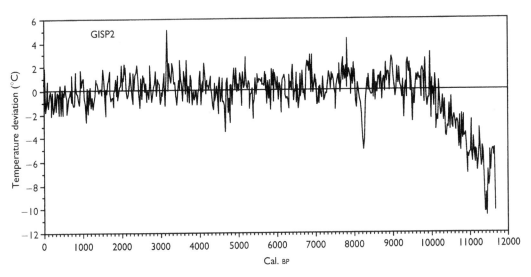

FIGURE 3.9 Holocene temperature variations recorded in the GISP2 ice core. (From Cuffey *et al.*, 1992, 1994)

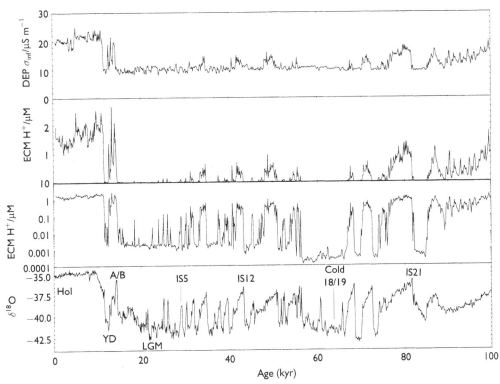

FIGURE 3.10 The electrical records and the oxygen isotope record for the last 100,000 years in the GRIP core. From the top: dielectric profiling (DEP) signal; electrical conductivity measurements (ECM) on a linear and on a logarithmic scale; and oxygen isotope signal. Climatic periods marked are the Holocene (Hol), Younger Dryas (YD), Allerød/Bølling (A/B), interstadial (IS) and cold periods between IS 18 and 19 (cold 18/19). (Adapted from Wolff *et al.*, 1997)

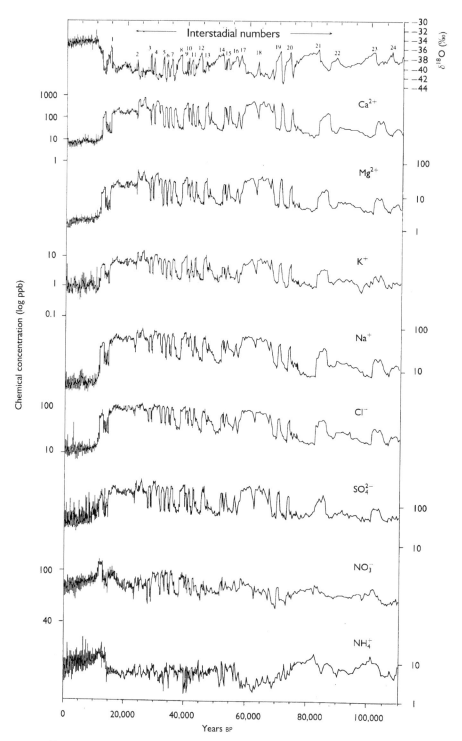

Figure 3.11 The $\delta^{18}O$ series and major ion series in the GISP2 ice core covering the last 110,000 years. (Adapted from Mayewski *et al.*, 1997)

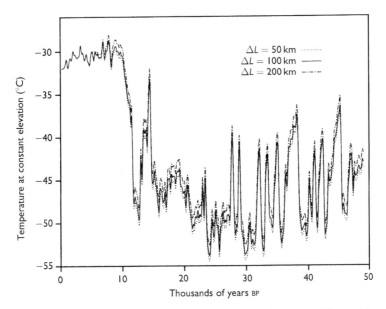

FIGURE 3.12 Temperature history from the GISP2 ice core according to calibrated isotope values and corrected for elevation changes. The data are smoothed with a 250-year triangular filter visualizing the effect of different elevation corrections. (Modified from Cuffey and Clow, 1997)

GISP2 ice cores. Volcanic eruptions and biomass burning events are also recorded by ECM.

Chemical analyses of sodium, potassium, ammonium, calcium, magnesium, sulphate, nitrate and chloride in the GISP2 core (Fig. 3.11) give a record of the atmospheric chemical composition and the history of atmospheric circulation in the mid-high latitudes of the northern hemisphere (Mayewski *et al.*, 1997). The record documents anthropogenic pollution, volcanic events, biomass burning, storminess over the oceans, continental aridity, and information related to the forcing of both high- and low-frequency climate events during the last 110,000 years (orbital cycles, Heinrich events and insolation variations).

Cuffey and Clow (1997) presented a model based on temperature, $\delta^{18}O$ variations, and a depth–age scale in the GISP2 ice core to obtain records of temperature, rate of accumulation, and the elevation of the ice sheet over the past 50,000 years (Fig. 3.12). Their model indicates that the temperature increased about 15°C from mean glacial to Holocene conditions. In addition, the average accumulation rate during the last glacial maximum

(15,000–30,000 yr BP) was around 25 per cent of the present accumulation rate, and long-term averaged accumulation rate and temperature correlate inversely over the last 7000 years. Interestingly, the Greenland ice sheet may have thickened during the glacial–deglacial transition, although the elevation history of the ice sheet is poorly constrained by the model. Studies of air content in the GRIP ice core (Raynaud *et al.*, 1997) seem to confirm a thickening of central Greenland during the Weichselian/Holocene transition. Figure 3.13 shows Holocene accumulation changes in the GISP2 ice core (Alley *et al.*, 1993). The frequency of melt layers in the GISP2 ice core decreased significantly from a maximum 7500–7000 yr BP (Fig. 3.14). Less frequent melt features after 7000 yr BP probably reflects a general summer temperature cooling (Alley and Anandakrishnan, 1995).

Dahl-Jensen *et al.* (1998) presented a 50,000-yr temperature history at GRIP and a 7000-yr history at Dye 3, using measured temperature profiles through the boreholes. The last glacial maximum, the Holocene climatic optimum, the Medieval period, the Little Ice Age, and a

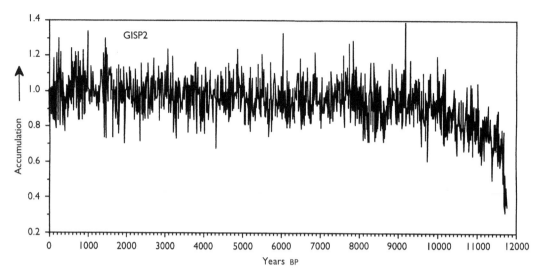

FIGURE 3.13 Holocene accumulation changes in the GISP2 ice core. (Alley *et al.*, 1993)

warm period around AD 1930 are recorded in the GRIP reconstruction, with amplitudes of −23, +2.5, +1, −1, and +0.5 Kelvin, respectively. The temperature in Dye 3 is similar to the GRIP history, but the amplitude is 1.5 times larger, suggesting greater climatic variability at that site.

The GRIP and GISP2 central Greenland ice cores provide evidence of abrupt climate changes during the last 100,000 years. Several of these variations have also been identified in deep-sea sediments from the North Atlantic. Steig *et al.* (1998) demonstrated that two of the most significant North Atlantic events (the

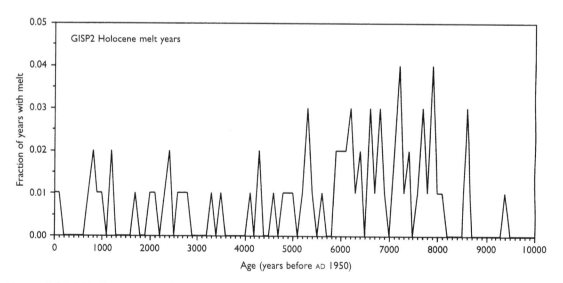

FIGURE 3.14 Melt layers according to age in the GISP2 ice core. The curve shows the 100-yr running mean of melt frequency (number of melt features per 100 years). (Modified from Alley and Ananda-krishnan, 1995)

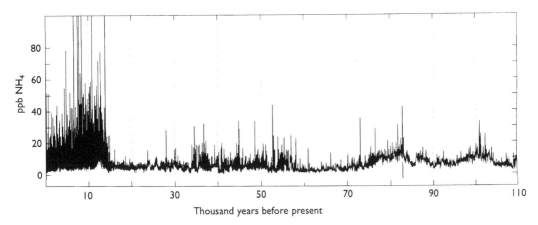

FIGURE 3.15 The GISP2 record of ammonium concentrations over the last 110,000 years. (Modified from Meeker *et al.*, 1997)

rapid warming marking the end of the last glacial period, and the Bølling/Allerød–Younger Dryas oscillation) are also recorded in an ice core from Taylor Dome, located in the western Ross Sea sector of Antarctica. The results from Taylor Dome contrast with data presented from ice cores in other regions of Antarctica, indicating asynchronous response between the northern and southern hemispheres.

A 110,000-year record of ammonium concentration in the GISP2 ice core (Fig. 3.15) has been used to infer terrestrial biological production and the pattern of atmospheric transport of ammonium (Meeker *et al.*, 1997). During warm periods, ammonium transport to Greenland was similar to the present. During extremely cold conditions, low ammonium levels in the ice core are the result of southerly excursions of the zonal polar circulation.

The completion of the ice drilling at the Vostok station in east Antarctica (78°S, 106°E, elevation 3488 m, mean annual temperature −55°C) in January 1998 (length of ice core 3623 m) has allowed the extension of the ice record of atmospheric composition and climate to the past four glacial–interglacial cycles (e.g. Petit *et al.*, 1999). The records from the Vostok ice core (Fig. 3.16) indicate that climate, within certain bounds, has almost always been in a state of change for the last 420,000

years. Features of the last glacial–interglacial cycle are also seen in previous cycles. During each of the last four glacial terminations, properties change in the following sequence: the temperature and atmospheric concentrations of CO_2 and CH_4 rise steadily, whereas the dust input decreases. During the last half of the temperature rise, CH_4 increases rapidly. The results from the Vostok ice core suggest climate forcing from orbital parameters followed by greenhouse gases and an albedo effect. The temperatures in Antarctica were higher during interglacials 5.5 and 9.3 (see Fig. 3.16) than during the Holocene or interglacial 7.5.

3.3 Gas content in ice cores

Below the surface of a glacier or ice sheet, air bubbles become isolated and trapped during the transformation of snow into ice. If no diffusion occurs, the bubbles contain 'fossil' air from the time of the inclusion. The composition of the bubbles in ice cores therefore provides a way to analyse changes in the atmospheric composition several thousands of years back in time. The reconstruction of atmospheric CO_2 concentration from ice cores is, however, not straightforward, due to (a) possible CO_2 production in the ice matrix as a result of chemical reactions

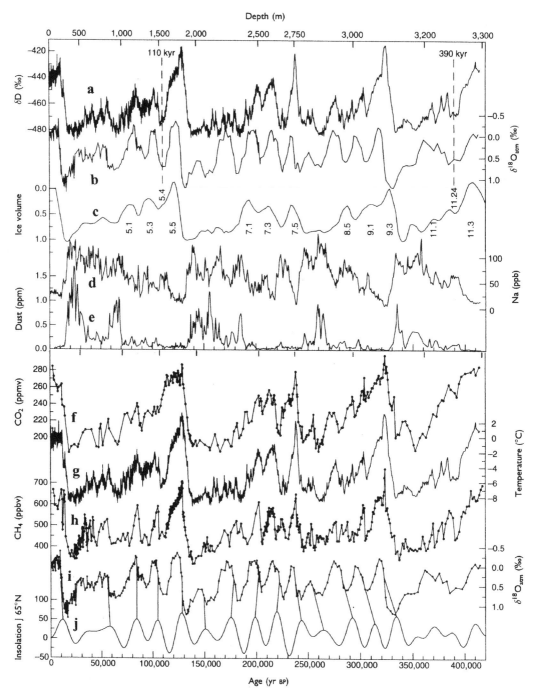

FIGURE 3.16 The Vostok ice core: depths on the top axis and time-scale on the lower axis with indication of the two fixed points at 110,000 and 390,000 yr BP. (a) Deuterium profile; (b) $\delta^{18}O_{atm}$ profile; (c) seawater $\delta^{18}O$ (ice volume proxy) and marine isotope stages; (d) sodium profile; (e) dust profile; (f) CO_2; (g) isotopic temperature of the atmosphere; (h) CH_4 (methane); (i) $\delta^{18}O_{atm}$; and (j) mid-June insolation at 65°N (in W m^{-2}). (Adapted from Petit *et al.*, 1999)

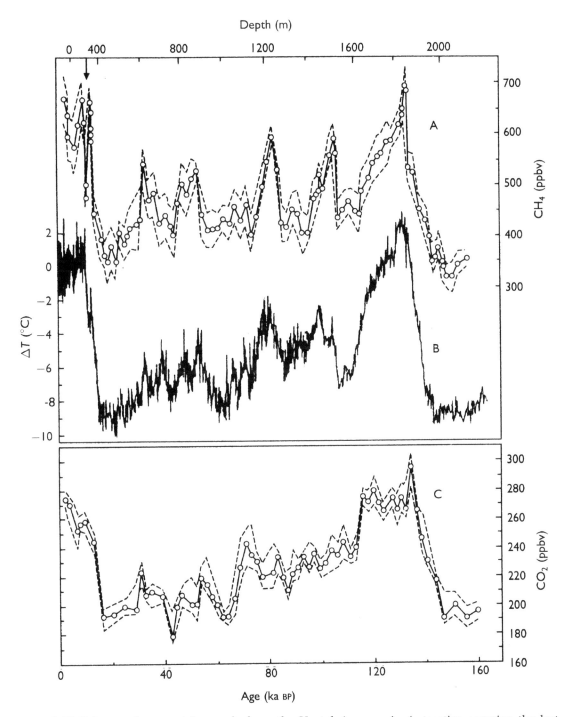

FIGURE 3.17 Palaeoenvironmental records from the Vostok ice core in Antarctica covering the last 160,000 years. (a) Methane content, mean values and 2-sigma uncertainty ranges; (b) surface palaeo-temperatures reconstructed from oxygen isotope variations; (c) variations in CO_2 content (average values and 2-sigma uncertainty ranges). (Adapted from Chappellaz *et al.*, 1990)

between impurities (Anklin *et al.*, 1995), and (b) fractionation of CO_2 between bubbles and clathrates (Miller, 1969).

Analyses of the air are made by placing a slice of the ice core in a vacuum chamber. Then the ice is cracked and the gases are analysed by chromatographic or spectrometric methods. In recent years, particular attention has been paid to the contents of the greenhouse gases CO_2 and CH_4.

To compare the rapid climate changes recorded in Greenland ice with the global trends in atmospheric CO_2 concentrations as recorded in Antarctic ice, an accurate common time-scale for the last glacial period was provided by Stauffer *et al.* (1998) using the records of global atmospheric methane concentrations from both the Greenland and Antarctic ice sheets. They found that the atmospheric concentration of CO_2 generally varied little with Dansgaard–Oeschger events, but varied significantly with Heinrich iceberg discharge events, especially those starting with a long Dansgaard–Oeschger event.

Records from both Greenland (Neftel *et al.*, 1982; Stauffer *et al.*, 1984) and Antarctica (Barnola *et al.*, 1987, 1991) ice sheets indicate that during the last ice age, atmospheric CO_2 levels were significantly lower than during other interglacials. Other atmospheric gases, such as CH_4 and nitrous oxide (N_2O) have also been found in lower concentrations in ice from the last glacial maximum (Chappellaz *et al.*, 1990; Raynaud *et al.*, 1992). During the transition from the last glacial to the Holocene (oxygen isotope stage 2/1), the atmospheric CO_2 levels increased from about 200 ppmv (parts per million by volume) to an interglacial value of approximately 280 ppmv. The CH_4 concentrations doubled from 350 to 650 ppbv, while N_2O values increased from around 190 ppbv to 270 ppbv (see Fig. 2.3).

The Antarctic Vostok ice core shows that during the last 160,000 years, CO_2, CH_4 and temperature (based on oxygen isotope data) have varied in phase (Fig. 3.17). Methane records from the Greenland ice cores (Dansgaard *et al.*, 1993; Blunier *et al.*, 1995) also demonstrate a close relationship with palaeotemperature records obtained from the ice cores.

The gas content in the upper ice core layers reflects human activity. A gradual increase in CO_2 in Greenland ice cores from the mid-eighteenth century and a significant increase during the last 50 years or so reflects the human impact on the atmospheric environment (Neftel *et al.*, 1982). Increases in fossil fuel combustion, clearance of forests and conversion of biomass to CO_2 have increased the level of atmospheric CO_2 by 20–30 per cent over the last 200 years. Recently, concentrations of chemical pollutants (for example lead, soot, nitrates) have increased significantly (Mayewski *et al.*, 1990).

A CO_2 record spanning the period between 40,000 and 8000 years BP was presented from the GRIP ice core (Anklin *et al.*, 1997). Ice-core records showed an increase in the atmospheric CO_2 concentration of 80–100 ppmv from the last glacial maximum to the early Holocene, from 200 ppmv to 290–310 ppmv (Fig. 3.18). Both the GRIP and the GISP2 ice cores show high CO_2 values in mild periods during the last glacial period. In Antarctica, however, the CO_2 ice-core records do not show similar high values. In addition, the early Holocene CO_2 values in the GRIP record are 20–30 ppmv higher than in the Antarctic records. The discrepancy between the GRIP and Antarctic CO_2 records was explained by Anklin *et al.* (1997) as autochthonous production of excess CO_2 by impurities in the ice. Since the carbonate concentration in Antarctic ice is less than in Greenland ice, the CO_2 records from Antarctica are considered more reliable than the Greenland CO_2 records. High-resolution records from Antarctic ice cores show that carbon dioxide concentrations increased by 80 to 100 ppmv, 600 ± 400 years after each warming of the last three deglaciation periods (Fischer *et al.*, 1999). High CO_2 concentrations can be sustained for thousands of years during glaciations despite strongly decreasing temperatures. The length of this phase lag is probably related to the duration of the preceding warm period, controlling the change in land coverage and build-up of the terrestrial biosphere.

A high-resolution Holocene record of CO_2 concentrations measured in air bubbles trapped in an ice core from Taylor Dome, Antarctica,

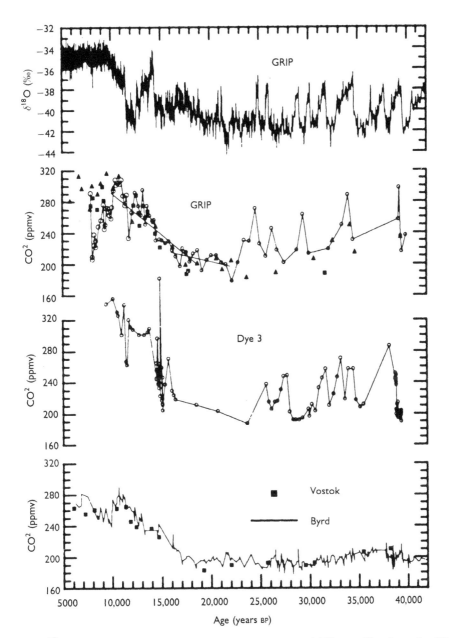

FIGURE 3.18 The $\delta^{18}O$ profile from the GRIP ice core versus age, and CO_2 profiles from the GRIP, Dye 3, Byrd and Vostok deep ice cores. (Modified from Anklin *et al.*, 1997)

shows a decrease of the CO_2 concentration from 268 ppmv at 10,500 yr BP to 260 ppmv at 8200 yr BP (Fig. 3.19). During the subsequent 7000 years, the CO_2 concentration increased almost linearly to 285 ppmv (Indermühle *et al.*, 1999).

3.4 Microparticles and radioactive tracers in ice cores

Evidence of several aspects of environmental variation have been obtained from ice cores.

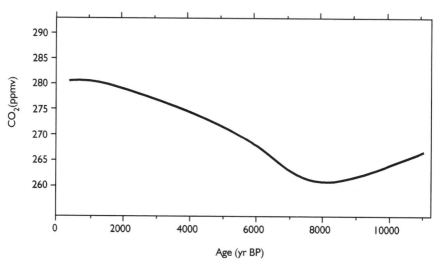

FIGURE 3.19 Holocene CO_2 curve from Taylor Dome, Antarctica. (Modified from Indermühle *et al.*, 1999)

Changes in atmospheric aerosol loading and the former extent of deserts or poorly vegetated areas have been detected using the dust content in ice cores (Jouzel *et al.*, 1990). Dust, defined as a suspension of solid particles in a gas or a deposit of such particles, is an important stratigraphic component in polar ice cores. Dust in the ice cores from Antarctica and Greenland is assumed to reflect the atmospheric dust burden at the time of deposition. Dust concentration in ice cores has been shown to vary seasonally. Annual dust cycles can thus help to identify annual layers of snow deposition, which can help to date the ice core. The presence of insoluble dust in polar ice means that annual insoluble dust maxima will be preserved and not affected by diffusion processes. Ram and Koenig (1997) presented a high-resolution dust profile from the GISP2 ice core through pre-Holocene ice, and also possibly penetrating Eemian ice, showing evidence of the Dansgaard–Oeschger events recorded in the oxygen isotope signal.

In the GRIP ice core, the mineralogical composition of atmospheric dust particles changes in phase with other palaeoatmospheric records (Maggi, 1997), indicating that climatic changes also affect the land surfaces of the dust source areas. Different mineralogical assemblages were systematically observed in cold and warm periods. During cold stages, quartz, illite, chlorite, micas and feldspars prevailed. This mineral assemblage is mainly related to mechanical weathering in arid and cold areas. Kaolinite and Fe (hydr)oxides dominated in warm periods, these minerals mainly being related to deep chemical weathering in warm and humid regions.

The size distribution of insoluble dust microparticles in the particle size interval 0.4–6 µm radius was measured in more than 1400 samples from the GRIP ice core (Steffensen, 1997), the total dust mass being strongly related to $\delta^{18}O$ variations. The volume distribution of the particles in the size interval 2.0–6.0 µm was almost the same in most analysed samples. Analysis of dust from the GISP2 ice core from the interval between 23,340 and 26,180 yr BP (around the last glacial maximum) shows that the probable source area was in eastern Asia, and not from mid-continental USA or the Sahara, which were suggested as potential source areas (Biscaye *et al.*, 1997).

The calcium (representing dust) concentration of the last 100,000 years from the GRIP ice core shows large variations (factor >100) (Fuhrer *et al.*, 1998). A significant change (factor 5–10) between low concentrations during the Dansgaard–Oeschger interstadial periods and high levels during colder periods is superimposed on a long-term trend with a further factor 5–10. Significantly higher wind velocities in the dust source area(s) in eastern Asia are probably required to explain these large and rapid increases in dust content. Changes in source area wind speed, almost entirely in phase with Greenland temperature changes, suggest that climatic parameters in high and low latitudes were strongly in phase during glacial periods, and that the atmospheric circulation system underwent almost simultaneous, large-scale variations during the last glacial period.

During the last glacial the dust content was about 200 times higher than at present (Thompson *et al.*, 1995). Variations in the concentration of sea-salt particles in ice cores are interpreted to reflect storminess (Shaw, 1989). Changes in sulphate in the Antarctic ice sheet have been used to indicate variations in biogenic productivity (Charleson *et al.*, 1987). Acidity profiles in ice cores provide records of sulphuric acid content, a proxy of volcanic aerosols and therefore the magnitude of former volcanic eruptions. Ice cores also contain radioactive fallout from atmospheric nuclear bomb tests since the late 1950s and early 1960s, and the Chernobyl reactor accident in 1986. Certain fallouts are useful stratigraphic marker horizons which can be used, for example, in mass balance studies.

Ion chromatography of Ca, Mg and Cl was used to study the cycle of marine and continental primary aerosols reaching Greenland (Angelis *et al.*, 1997). Calcium is a good indicator of input from continental sources. The calcium content of continental background aerosols over Greenland was considerably higher during the glacial period. The marine Na component reflects sea-salt aerosols, showing an inverse linear relationship with $\delta^{18}O$. The clorine to marine sodium weight ratio increases with temperature, from values close to the bulk seawater ratio during the last glacial maximum to significantly higher values during the Holocene.

A nearly continuous record of ^{10}Be (half-life of 1.5×10^6 years) concentrations was reported from the GISP2 ice core, spanning the interval between 3288 and 40,055 years BP (Finkel and Nishiizumi, 1997). ^{10}Be concentrations show a strong correlation with $\delta^{18}O$ and a more weak correlation with snow accumulation rate. There is a strong correlation between $\delta^{14}C$ and ^{10}Be on a centennial time-scale.

Natural cosmogenic ^{36}Cl is produced in the atmosphere mainly by neutron- and proton-induced spallation reactions on ^{40}Ar. The production rate of ^{36}Cl is proportional to the cosmic ray flux, which is modulated by the Earth's magnetic field and the magnetic properties of the solar wind. Within approximately two years, ^{36}Cl is removed from the atmosphere and stored in terrestrial deposits such as ice sheets (e.g. Baumgartner *et al.*, 1997). Except for the Holocene/Weichselian–Wisconsin transition, the GRIP ice core was analysed continuously for the cosmogenic isotopes ^{10}Be, ^{36}Cl and ^{26}Al.

3.5 Volcanism and climate

Large, explosive volcanic eruptions commonly inject huge amounts of silicate microparticles and acid gases into the stratosphere. This normally leads to warming of the stratosphere due to absorption of incoming solar radiation. Sulphate aerosols are formed, causing a cooling of the lower troposphere by back-scattering of solar radiation (e.g. Rampino and Self, 1982). The volcanic particles have a residence time in the stratosphere of at least one to three years and are distributed by stratospheric winds to form a veil over much of the planet. This aerosol veil prevents the passage of incoming solar radiation to the Earth's surface, causing a lowering of surface temperatures.

The Greenland Ice Sheet Project 2 (GISP2) and the Greenland Ice Core Project (GRIP) from Summit, Greenland, provide records of aerosol (H_2SO_4) and tephra particles from past volcanic activity (Zielinski *et al.*, 1997). These continuous records are helpful in producing a hemispheric and global chronology of explosive volcanism and assessing the climatic effects of volcanism. The volcanic SO_4^{2-} records for the last 110,000 years show a strong connection between periods of

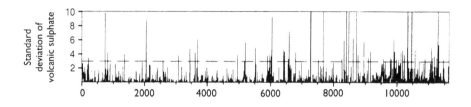

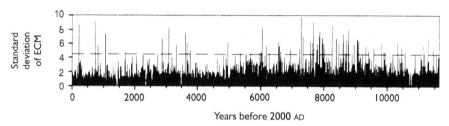

Years before 2000 AD

FIGURE 3.20 The Holocene record of electric conductivity measurements (ECM) and volcanic sulphate in the GISP2 ice core. The records are presented as standard deviation units. Adapted from Taylor *et al.* (1997) and Zielinski *et al.* (1997)

enhanced volcanism and periods of climate change. An increasing number of explosive volcanic eruptions at 27,000–36,000 and 79,000–85,000 years BP may reflect initial ice sheet growth. The largest number of volcanic events occurred during the period between 17,000 and 6000 years BP, probably reflecting increased crucial stress due to variations in ice-sheet loading and volume changes of ocean water (Sejrup *et al.*, 1989). The Holocene records of electric conductivity measurements (ECM) and volcanic sulphate in the GISP2 ice core are shown in Figure 3.20.

The majority of the largest volcanic eruptions have been in South East Asia: Tambora in AD 1815, Krakatoa in AD 1883, Agung in AD 1963, and Pinatubo in AD 1991. Normally, the climatic effect of a volcanic eruption is greatest the following year. The 'year without summer' in 1816 is believed to have resulted from the 1815 Tambora eruption.

It has been proposed that volcanism has had a profound effect on climate and glacier variations during the last millennium (Lamb, 1970; Baldwin *et al.*, 1976; Mass and Schneider, 1977; Bradley, 1978, 1988; Miles and Gilder-sleeves, 1978; Hammer *et al.*, 1980; Porter 1981a, 1986; Self *et al.*, 1981; Gilliland, 1982;

Rampino and Self, 1982; Kelly and Sear, 1984; LaMarche and Hirschboeck, 1984; Sear *et al.*, 1987; Oerlemans, 1988; Baille, 1989; Scuderi, 1990). Porter (1986) discussed the pattern and forcing of northern hemisphere glacier variations during the last millennium and found a close relationship between glacier fluctuations and variations in Greenland ice core acidity (Hammer *et al.*, 1980), indicating that sulphur-rich aerosols were a primary forcing mechanism of recent glacier fluctuations. Local and regional temperature drops related to the largest eruptions were calculated to 0.5–1.2°C. By keeping (winter) precipitation constant, this corresponds to a depression of glacier equilibrium line altitudes in the range of 80–200 m, which coincides with values found for the culmination of the Little Ice Age (Porter, 1986; Oerlemans, 1988).

3.6 Mapping and measuring glacier-front variations

Long-term glacier observations which were coordinated internationally began in 1894 with the establishment of the *International Glacier Commission* in Zurich, Switzerland.

The goal of this worldwide monitoring programme was to provide information on mechanisms of modern climate and glacier variations. At present, glacier variations are recognized as summer temperature and winter precipitation indicators used in the early detection of possible human-induced climate change (IPCC, 1992, 1995).

Glacier fluctuations contribute information about natural climate variability and rates of change with respect to short- and long-term energy fluxes at the glacier surface. Historical and Holocene glacier fluctuations, reconstructed from direct measurements, paintings, written sources and moraines, indicate that the glacier extent in many mountain regions has fluctuated considerably over the past centuries and millennia. The range of variability is defined by the early Holocene climate optimum and today's reduced stages, and the maximum Little Ice Age glacier extent.

Glacier margins advance or retreat, with variable time lags, in response to variations in *glacier mass balance*. *Ablation* removes ice from the glacier and the *horizontal velocity component* carries ice forward. Boulton (1986) demonstrated that a glacier margin will remain in the same position when the horizontal velocity component is equal to the horizontal component of ablation. Although the frontal position is stationary, the ice is in motion, but is removed from the glacier at a rate equal to the velocity. Frontal retreat takes place when the horizontal velocity component is less than the horizontal ablation component, whereas glacier advance occurs when the horizontal velocity component is larger than the horizontal ablation component. During the winter season, glacier sliding velocities tend to be low due to a lack of lubricating meltwater at the glacier base. Because ablation rates are generally negligible during the winter compared with the summer, however, the relative magnitudes of the horizontal velocity component and the ablation rate over the year tend to cause winter advance and summer retreat. Commonly, winter advances start late in the ablation season when melting at the margin does not exceed the forward flow of glacier ice. Normally, the horizontal ablation

component is low in late winter, causing the small winter flow velocities to produce small glacier advances. Despite higher summer flow velocities than in winter, high summer ablation rates cause net retreat of the glacier (e.g. Benn and Evans, 1998).

On longer time-scales, glacier fronts are subject to advance and retreat as a result of climate change or *internal instabilities*. Climatic influences on the frontal response can be divided into factors causing changes in ablation and accumulation. Debris-covered glacier fronts are, however, rather insensitive to changes in mass balance. Varying amounts of ice flowing through a glacier causes changes in ice thickness and gradients and thereby influences the *driving stresses*. Advance and retreat of the glacier front normally lags behind the climate forcing because the signal must be transferred from the accumulation area to the snout. This is referred to as the *time lag* or the *response time*, which is longest for long, low-gradient and slow-moving glaciers, and shortest on short, steep and fast-flowing glaciers (e.g. Johannesson *et al.*, 1989; Paterson, 1994). *Kinematic wave* theory has been applied to calculating response times (Nye, 1960; Paterson, 1994). However, physically-based *flow models* may help to determine the response times more precisely (van de Wal and Oerlemans, 1995).

3.7 Terminal moraines

Present and former positions of glaciers are marked by different moraine types formed by the deposition of sediment at the margins, or by stresses induced by the glaciers. Such deposits exhibit a number of features, such as glacitectonic landforms, push and squeeze moraines, dump moraines, and later/frontal fans and ramps. It is often difficult to classify ice marginal deposits. Moraines of supraglacial and englacial origin are difficult to recognize because the material is lowered on to the ground during the retreat of the glacier margin. The outer moraine ridge formed at the limit of the glacier advance is commonly termed the terminal moraine, while younger moraines within the terminal moraine are called recessional moraines. Recessional moraines form during minor advances or stand-stills during general retreat. Terminal and recessional moraines my be subdivided into frontal and lateral parts, or latero-frontal moraines. Recessional moraines formed on a yearly basis are termed *annual moraines*.

The ability of glaciers to deform bedrock and sediments into thrust moraines and sheets has been known since early Quaternary studies from Sweden, Denmark, Germany, Poland, the UK and North America (e.g. Aber *et al.*, 1989). Four basic types of glacitectonic landforms are recognized: (a) hill–hole pairs; (b) composite ridges and thrust block moraines; (c) cupola hills; and (d) megablocks and rafts.

Push moraines are normally small moraine ridges (usually less than 10 m high) formed during minor glacier advances. These moraines can form either at subaqueous or terrestrial ice margins. Cross-sections of push moraines are often asymmetric with gentle proximal and steep distal slopes. Push moraines are broadly arcuate in form, but small-scale morphological features reflect the morphology of the glacier margin. At glacier margins with radial crevasses, push moraines have a characteristic saw-tooth shape.

Material accumulating on the glacier surface is subject to remobilization by mass flow, sliding, falls or fluvial transport. The remobilization may result in dumping of the material during the glacier recession. *Dump moraines* therefore form where the ice margin remains stationary during debris accumulation (Boulton and Eyles, 1979).

Debris flows and glaciofluvial processes around stationary glacier margins may form *latero-frontal fans* and *ramps*. The distal slopes of such forms have lower gradients than dump moraines. The proximal slopes are, however, steep, reflecting the former ice-contact face.

3.7.1 Dating techniques in glacier forelands

The resolution and reliability of any moraine chronology is only as good as the precision and accuracy of the dating technique, and the interpretation of the relationship between dated layers and geomorphological processes. The interpretation of dated chronological records obtained from landforms, and moraine sequences in particular, rarely pay attention to the inherited complexity in interpreting landform formation. The uncertainties may include availability of datable material, the sensitivity of the system to change, and the subsequent erosion or destruction of the landform (Kirkbride and Brazier, 1998).

Dating techniques found useful for retreating glaciers are normally divided into four categories: (1) observation and measurements; (2) historical documents; (3) biological dating; and (4) physicochemical techniques. In the investigation of Holocene glacier and climate variations, where precise dates are needed for frequent and low-amplitude events, it is important to adopt a critical approach to dating (e.g. Matthews, 1985, 1997). Problems related to historical dating may be the small number of glaciers for which there are reliable historical evidence, the difficulties of dating moraines by the use of historical data, and the lack of reliable data for the period prior to the Little Ice Age maximum. Lichenometry may be limited by the relatively extensive Little Ice Age glacier advances in some areas, the sparse independent historical evidence for calibration of lichenometric dating curves in the older parts, and the reliance in some regions on radiocarbon-dated palaeosols for calibration of lichenometric growth curves. Relative dating based on weathering-based criteria (e.g. Schmidt hammer, surface roughness and weathering rinds) may be limited by the lack of quantitative data on weathering rates and reliable calibration points. With radiocarbon dating, major problems include the reliance of moraine dating of palaeosols, the existence of steep age–depth gradients in buried soils, the inherent precision of radiocarbon dates (one or two sigma), and finally, problems in interpreting various stratigraphical contexts, for example, glacial, glaciofluvial and glaciolacustrine sediments (Matthews, 1997).

3.7.1.1 Observation and measurements

Glacier observation and measurements for scientific purposes began in the European Alps. In 1842 a map of the Unteraar glacier was constructed (scale 1:10,000) by M.J. Wild (Zumbühl *et al.*, 1981). Another and cheaper approach to mapping glacier snouts is to measure the distance from one or more fixed marks in front of the glacier. In Switzerland, between 50 and 100 glaciers have been measured since the year 1890. These frontal measurements are supplemented by terrestrial and air photographs, and most recently by satellite imagery. In Norway, glacier-front positions have been measured systematically from about AD 1900.

3.7.1.2 Historical documents

Historical documents have been considered to be one of the most accurate sources for reconstructing recent glacier variations. In addition, this information has been used to calibrate data on glacier variations further back in time. The Icelandic Sagas (AD 870–1264) seem to be the oldest documents mentioning glacier variations. Some of the oldest historical data do not, however, fulfil modern scientific standards and must not be looked upon uncritically.

In Iceland, Norway and the Alps, agricultural land was abandoned due to expanding glacier fronts during the Little Ice Age (e.g. Grove, 1988). Around Jostedalsbreen in western Norway, historical evidence shows that the advances of Nigardsbreen in Jostedalen and Brenndalsbreen in Olden caused the most severe damage, and that which affected the farm Tungøyane in Olden was the most severe. The destruction of Tungøyane took place over a period of about 40 years when the glacier front of Brenndalsbreen was situated in the vicinity of the farm, causing a series of avalanches and floods over the farmland. On 12 December 1743 the farm was totally destroyed by an avalanche from the glacier, and all but two persons on the farm were killed (Nesje, 1994). Information about the Little Ice Age glacier damage in Norway has been obtained through records of tax reductions (for further details, see Grove and Battagel, 1983; Nesje, 1994).

From the seventeenth century onwards, several persons visiting the glaciers left paintings, drawings and photographs providing material for reconstruction of glacier positions and later fluctuations. In the Swiss Alps, for example, the Lower Grindelwald Glacier has 323 illustrations to document its former extent, and together with written evidence, this forms the basis for a detailed reconstruction of the glacier back to AD 1590.

3.7.1.3 Biological dating

Two biological dating techniques have proved useful for dating glacier forelands: dendrochronology and lichenometry. In regions where glaciers descend into areas with trees, the annual pattern of tree growth may be affected by the proximity to the glacier. In recently deglaciated glacier forelands, it is important to establish the age of living trees by counting annual rings and the age of abnormal (normally reduced) growth rates both in living and dead trees (e.g. Schweingruber, 1988). The age of the oldest living tree provides a minimum age for deglaciation. This technique has been used with success, especially in western North America (e.g. Luckman, 1986). One dendrochronological technique tries to date glacier-induced growth rates from trees partly broken or tilted but not overrun or killed by the glacier (e.g. Luckman, 1986). Trees killed by the glacier and later exposed by glacier retreat may be cross-dated with living trees or dated by the radiocarbon method, an approach widely used in the European Alps (e.g. Holzhauser, 1984). In front of Briksdalsbreen, two *Salix* trunks, exposed during recent glacier advance, were dated to 7650 ± 85 radiocarbon yr BP (8405 (8485–8340) cal yr BP) and 7530 ± 100 radiocarbon yr BP (8325 (8400–8165) cal yr BP) (Nesje, unpublished).

Lichenometric dating, developed in the context of recently deglaciated terrain by Beschel (1950, 1957, 1961), has been widely used, in particular the yellow-green *Rhizocarpon geographicum* (Innes, 1985a,b, 1986a,b). There are two main applications of the method. The 'indirect' approach is based on the assumption that there is a relationship between lichen size and terrain age. Interpolation between points of known age can be used to date other surfaces by using lichen size. The greatest limitation of the indirect lichenometric approach is the need for several surfaces of known age (control points). Commonly, the age estimates obtained by this method are given with an accuracy of ±10 per cent (Bickerton and Matthews, 1993). McCarroll (1994) developed a new approach to lichenometric dating, based on large samples of the single largest lichen on each boulder. On surfaces of uniform age, lichen sizes are close to normally distributed, and mean values can therefore be used to construct lichenometric dating curves.

Direct lichenometric dating means that a growth curve is established by direct measurements of lichen growth rates. This dating technique has, however, had little success, mainly due to slow lichen growth, great variability of lichen growth rates, and problems of linking growth curves to site age (Matthews, 1992).

3.7.1.4 Physicochemical techniques

Several physicochemical dating techniques based on isotopes or radiological processes, rock surface weathering, tephrochronology, varves (annual laminae in sediments) and palaeomagnetism have been applied to glacier forelands. Radiocarbon dates for wood, peat, soil or lacustrine sediments obtained from glacial forelands can either be of the actual event (contemporary), bracketing or limiting (Porter, 1981b). Most commonly, samples provide either maximum or minimum ages of the actual event. Dates from below and above may bracket the age of the deposited layer.

Radiocarbon dating of different soil fractions from palaeosols has provided information on glacier advance and retreat. This approach has been used with great success in New Zealand, the European Alps and Scandinavia. Dating of thicker soils developed over a considerable time span is somewhat problematic, because of problems in isolating organic layers/fractions of different ages due to greater mean residence time (Matthews, 1985). In

southern Norway, Matthews and Dresser (1983) and Matthews and Caseldine (1987) reported steep age/depth gradients in soils beneath Little Ice Age moraines, ranging from about 4000 yr BP at the bottom to several hundred years at the top. The radiocarbon method itself poses a problem with precision. Age limits given with two standard deviations (95 per cent certainty) normally give an age uncertainty in the order of ±100 years. When calibrating radiocarbon dates into calendar ages, dates younger than about 400 years give equivocal dates. As an example, a radiocarbon age of 220 ± 50 radiocarbon yr BP (1 sigma) is equivalent to calendar age ranges of 150–210, 280–320 and 410–420 calibrated yr BP (Porter, 1981b).

Tephra layers are stratigraphical marker horizons which may indicate the relative age of the overlying or underlying deposits. Tephra layers have been used to date moraines in areas subject to volcanic eruptions, particularly in Iceland (e.g. Dugmore, 1989) and North America (Porter, 1981b).

Laminae/varves in proglacial lakes may indicate upstream glacier fluctuations (e.g. Karlén, 1976, 1981; Nesje et al., 1991; Karlén and Matthews, 1992; Matthews and Karlén, 1992; Dahl and Nesje, 1994, 1996). (For further details, see Section 3.8.)

Rock surface colour (Mahaney, 1987), rock disintegration (Innes, 1984), rock surface hardness and roughness (Matthews and Shakesby, 1984; McCarroll, 1989; McCarroll and Nesje, 1993) and weathering-rind thickness have been used to obtain relative ages. Theoretically, weathering rates or degree of rock surface weathering can be calibrated with other dating techniques, for example the radiocarbon method, to get absolute dates (e.g. Colman, 1981).

In conclusion, direct observation and measurements, historical documents, dendrochronology, lichenometry, radiocarbon dating and tephrochronology are considered the most accurate dating approaches in glacier forelands. If possible, multi-parameter methods or several dating techniques should be used to obtain the most accurate terrain age in glacier forelands.

3.8 Lacustrine sediments

One of the best archives of terrestrial palaeo-climate information is lake sediment. Annual and decadal climate shifts influence sediment production and deposition in lakes. During the winter, lakes, at least in mountainous regions, are normally frozen and little clastic sediment enters the lakes. Organic production is effectively reduced due to little sunlight and snow-covered lake ice. In spring and early summer, strengthened sunlight, ice-free lake conditions, and nutrients released from melting ice and snow, support diatom blooms. Pollen from adjacent plants is also deposited in lakes during the summer season. Terrigenous silts and clays, deposited in the lakes from surface runoff from rainfall and snow-melt throughout the summer, settle out of suspension and form another sediment layer. This seasonal alternation of depositional regimes results in the annual production of laminae/varves.

Sediments accumulating in proglacial lakes contain information about glacier fluctuations in the form of variations in particle size, sediment thickness, and organic minerogenic content (e.g. Karlén, 1976, 1981, 1988; Leonard, 1986; Nesje et al., 1991; Karlén and Matthews, 1992; Matthews and Karlén, 1992). Such sediments commonly form continuous records reflecting climate and glacier fluctuations, because the relative amount of minerogenic silt and clay eroded by the glacier varies with glacier activity. Glacier fluctuations may therefore be compared directly with variations in pollen influx and the relative pollen content, which reflect local vegetation changes. In lakes with rhythmic sedimentation, precise dating may provide information about climatic/environmental changes and different response times between physical and biological systems.

Several factors, including the size, depth and bathymetry of the lake, the altitude of the lake, its distance from the glacier, the proportion of the catchment glacierized, whether coarse sediments are trapped in upstream lakes, and the form of the surrounding hill slopes and especially their exposure to avalanche activity, must be seriously evaluated in lacustrine sediment studies. Lacustrine sediment studies normally include: loss-on-ignition (LOI) measurements, grain-size distribution (sedigraph), X-radiography of sediment cores, visual counting of laminae/varves, ^{210}Pb and accelerator mass spectrometry (AMS) radiocarbon dating, palaeomagnetic measurements, pollen counts, diatoms, plant macrofossils and chironomids.

3.8.1 Laminae/varves

Palaeoenvironmental interpretations and reconstructions based on varved sediments began early in the history of varve-sediment research, and focused on using the interannual variations in the varve laminae thickness as a monitor of past environmental change (Anderson, 1961; Saarnisto, 1979; Renberg and Segerström, 1981; Perkins and Sims, 1983; Leonard, 1986; Østrem and Olsen, 1987; Cromack, 1991; Lotter, 1991; Deslodges, 1994). Variations in varve/laminae thickness have since been related to other environmental variables, such as temperature, precipitation, wind stress, coastal upwelling, and glacier activity/run-off. Palaeoenvironmental reconstructions have been derived from the coupling of varve chronologies with data from other evidence (pollen, diatoms, plankton, palaeomagnetic variations and geochemical changes).

The application of annually laminated (varved) sediments in palaeoenvironmental research has expanded with the growing need for millennia-long records of interannual to century-scale climate variability. In cases where the annual nature of the varves can be confirmed by radiometric and other methods, the laminae/varves provide annual resolution records that can be coupled with a wide range of palaeoenvironmental proxies to provide time series with accurate seasonal- to centennial-scale sample resolution. Proper interpretations require that the climate response time of the signal is taken into account.

Palaeoclimatic research has increasingly relied on the use of varved (annually laminated) lake and marine sediments to provide high-resolution chronologies for climate reconstructions.

Sites with laminated/varved sediments have complemented those with other sources of palaeoclimatic information, such as tree-rings, corals, ice cores and historical documents (Bradley and Jones, 1993). Where other annually dated records of palaeoenvironmental variability exist, varved sediments can provide useful complementary or longer time series. Although varves by definition are annual in nature, there will commonly be some significant counting error associated with varve-based chronologies. In addition, climate interpretations based on varved sediments are also subject to some controversy. Defining a clear, unambiguous palaeoenvironmental signal recorded in the sediments may be difficult. Of similar importance is the need to establish the climatic response time of the recorded signal.

Although varve-based palaeoenvironmental reconstructions are subject to some inherent uncertainties, they may ultimately provide some of the longest and most useful records of past annual to decadal-scale climatic variability. In contrast to most proxy time-series of annual/decadal climatic variability, varved sediments span millennia. Laminated/varved sediments therefore provide a key resource for unravelling the range of climatic variability, and for understanding how this past variability may be effected by climatic forcing unlike that of the present.

3.8.1.1 Environments of lamina formation and preservation

There are two fundamental requirements for the development of laminated sediment sequences:

(1) Variation in input/chemical conditions/ biological activity that will result in compositional changes in the sediment.
(2) Environmental conditions that will preserve the laminated sediment fabric from bioturbation. In lakes, strong seasonal signals are dominant while preservation is effected by bottom-water anoxia resulting from stratification, and high sedimentation rates.

Varves are defined as laminae or group of laminae interpreted to represent one year's deposition. Lamination can be formed by changes in terrigenous sediment grain-size, differences in biogenic influx and/or production, seasonal diatom production cycles, chemical variations, and water-column precipitation.

3.8.1.2 Techniques for the study of laminated sediments

The most common laboratory techniques used to investigate laminated sediments include X-radiography, core surface photography, digital imaging, optical and scanning electron microscopy.

X-radiography. Briefly, X-radiography is based on the differential passage of X-rays through a heterogeneous media, on to X-ray-sensitive photographic film. Dense, minerogenic laminae will produce light negative images and more penetrable, biogenic laminae will produce darker negative images. Thus X-radiography provides useful information on the broad structure of the sediment column.

Digital imagery. This is a relatively new technique for producing core photographs. In digital imagery, a digitizer scans the core surface and the colour/grey scale of the sediment is recorded, producing an image similar to the standard photograph.

Optical microscopy. This technique has been used to analyse sediment fabric elements such as microfossils and lamina boundaries (i.e. whether they are gradational or sharp).

Scanning electron microscopy. The scanning electron microscope (SEM) is an ideal tool for high-resolution analysis of laminated sediments. Preparation of unconsolidated sediment using fluid-displacive resin embedding prevents fabric disturbance, and produces high-quality thin sections.

3.8.1.3 Establishing a sediment chronology from laminated/varved sediments

Using varved sediments in palaeoenvironmental reconstruction begins with the establishment of a firm chronology, included a dating uncertainty. Several of the potential problems

Box 3.1 Extracting the climate signal from varved sediments

Varved sediments have been used to generate palaeoenvironmental reconstructions based on variations in the varve thickness and composition, as well as on variations of fossil or other signals in the sediments. However, it may be difficult to identify the type of environmental signal recorded in the sediments, and to quantify the relationship between sediment variations and a given environmental parameter (temperature, precipitation, runoff). It can be hard to estimate the rate at which a sediment-related change responds to a given environmental parameter. Sometimes, it has been difficult to quantify the varve–climate relationship, which can result from: (1) the lack of suitably long instrumental time series of appropriate environmental data near sites with laminated sediments; (2) an inaccurate varve age model; (3) the complex multivariate or evolving nature of the sediment–environment link; (4) the dominance of sedimentological/environmental noise over environmental signal; and (5) a slow or uneven rate of response to environmental change.

The environmental response times of many potential palaeoenvironmental signal-carriers are rapid enough to eliminate concern. Most biotic and abiotic factors respond almost immediately to environmental changes, and carry this signal quickly to the sediment. Varved sediments are by definition not significantly bioturbated, but it is possible for periodic downslope sediment movement to occur.

in building a varve-based chronology are analogous to those dealt with in dendrochronology. The first step in generating a varved-sediment record is to obtain undisturbed samples of the sediment. For most applications, it is important to obtain the uppermost sediments, including the sediment–water interface. This allows the varve counts to be placed in an exact calendar-age framework.

A well-dated varved sediment record requires multiple cores. Single cores are subject to many core-specific interpretational problems which can be minimized using comparisons between individual cores. These problems commonly include core gaps, localized core disturbance, missing varves, or an anomalous number of laminae deposited per year. In practice, the identification of varves is subjective due to sedimentological noise, which can be reduced using comparisons between multiple cores. In most cases, it is unlikely that the varve-based ages are determined without error, and errors in the order of 5–10 per cent should be assumed, unless a lower error estimate can be justified. An essential first step in using laminated sediments for geochronology and palaeoclimatic reconstruction is to confirm that the laminations are truly annual (varved). This can be accomplished by varve-dating known stratigraphic events, by radiocarbon dating (calendar dates), and by bomb-nuclide profiles. ^{210}Pb is one way to confirm that the varves are annual. The temporal resolution of radiocarbon dates is, however, too coarse to be ideal for confirming varve chronologies, but can help to date sediments older than 150 years.

3.9 Marine sediments

In the oceans, sediments have accumulated for millions of years. The sediments consist partly of terrigenous material, which is detritus derived from erosion of the surrounding land masses, and biogenic material consisting of calcareous and siliceous skeletal fragments from micro-organisms. The terrigenous material is brought to the ocean floor mainly by turbidity currents, bottom currents, wind and ice. In the mid- and high-latitude oceans, Ruddiman et al. (1989) found that terrigenous detritus was mainly deposited during glacial episodes,

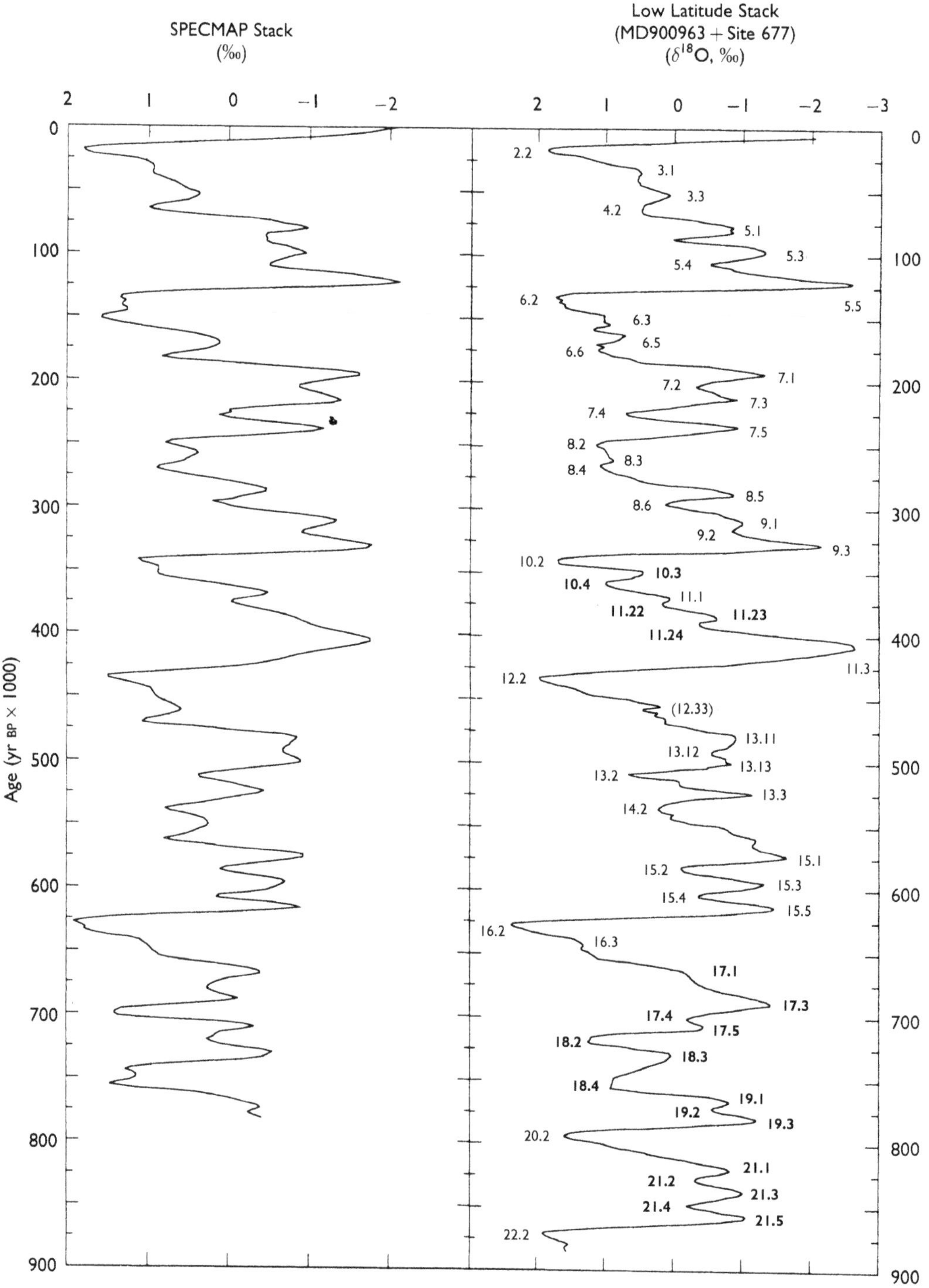

FIGURE 3.21 The SPECMAP stack plotted against age using the time-scale developed by Imbrie *et al.* (1984) (left) and the low-latitude stack tuned to the ice volume prediction model of Imbrie and Imbrie (1980) with numbering of the isotopic events. (From Bassinot *et al.*, 1994)

reflecting transport by means of ice rafting of glacially eroded debris. In ocean sediment records, the ice-rafted debris (IRD) reflects former glacial episodes. In the North Atlantic, IRD deposition has been the most important mechanism for terrigenous sediment transport into the oceans; at 45°N, cycles of IRD deposition during the last glacial period have been recognized. These have been termed Heinrich layers (Heinrich, 1988) and reflect episodes of IRD arrival from icebergs drifting eastwards from the Laurentide ice sheet margins (Alley and MacAyeal, 1994).

In the deep oceans, the sediments are normally finer grained and dominated by biogenic material of carbonaceous and siliceous particles from marine micro-organisms (marine oozes). The detailed data for environmental change in the oceans come from the chemical and isotopic composition of these marine organisms. Variations in aluminium, barium, calcium and cadmium, and the isotopes of carbon, oxygen and uranium, reflect changes in circulation, nutrient supply and water temperature in the oceans. The application of oxygen isotope analysis to deep-ocean sediments has revolutionized Quaternary science (Figs 3.1 and 3.21). Oxygen exists in three isotopes (^{16}O, ^{17}O and ^{18}O), and ^{16}O and ^{18}O are used for oxygen isotope analysis of marine deposits. Ratios between ^{18}O and ^{16}O vary between 1:495 and 1:515 (average of about 1:500 or 0.2 per cent ^{18}O). Ratios between the two oxygen isotopes are measured in relative deviations ($\delta^{18}O‰$) from a laboratory standard value. The common standards used are *PDB* (belemnite shell) for carbonate analysis, and *SMOW* (*Standard Mean Ocean Water*) for analysis of water, snow and ice. PDB is +0.2‰ in relation to SMOW.

4

Glacier dynamics

4.0 Chapter summary

This chapter gives an overview of the present distribution of glaciers, the physical properties of glacier ice, and the classification of glaciers and ice sheets. It also explains the temperature distribution in glaciers and ice sheets. The importance of glacier monitoring, lately by satellites, is assessed. Furthermore, determination and reconstruction of the equilibrium line altitude (ELA), mass balance on glaciers, and regional, long-term mass balance variations are dealt with. Chapter 4 explains frontal glacier variations, the terms response time/timelag, how glaciers move, and gives examples of supraglacial features on glaciers. In addition, the chapter deals with glacier hydrology, calving glaciers, surging and tide-water glaciers, and finally, how ice-surface profiles are reconstructed and how basal shear stress is calculated.

4.1 Present distribution of glaciers

The large ice sheets of Antarctica (85.7 per cent) and Greenland (10.9 per cent) together represent 96.6 per cent of the world's total glacierized area (13,586,310 km^2). About two-thirds of the remaining 3.4 per cent (ca. 550,000 km^2) are high-latitude ice caps and ice fields, and one-third mountain glaciers. Table 4.1 shows the distribution of glacierized areas of the world. The volumes of the Antarctic and Greenland ice sheets are estimated to be 30,100,000 km^3 (91.8 per cent) and 2,600,000 km^3 (7.9 per cent),

respectively. The volume of the remaining glaciers is estimated to 100,000 km^3 (0.3 per cent), making a total glacial volume of 32,800,000 km^3. Estimates suggest that about 80 per cent of the world's freshwater is stored in glacier ice. If the West and East Antarctic ice sheets melted, sea-level would rise by 5 m and 60 m, respectively. Greenland would add another 5 m to sea-level rise, while the remaining glaciers would contribute less than 1 metre. Thus, if all glaciers and ice sheets melted, global sea-level would rise about 70 m.

4.2 Glacier types

A classification of glaciers can either be done on the basis of glacier *morphology* or the *physical* properties.

4.2.1 Morphological classification

Glacier form is a function of climate and topography, and therefore a variety of glacier morphologies exists, from the smallest niche glacier to the largest ice sheet. A first-order classification into ice sheets and ice caps unconstrained by topography can be subdivided into a second-order classification of ice domes, ice streams and outlet glaciers. Glaciers constrained or controlled by topography include icefields, valley glaciers, transection glaciers, cirque glaciers, piedmont lobes, niche glaciers, glacierets, ice aprons and ice fringes. Marine glaciers include ice rises, glacier ice shelves and sea-ice ice shelves (Ommanney, 1969;

TABLE 4.1 Distribution of glacierized areas of the world (adapted from World Glacier Monitoring Service)

Continent	Region	Area (km^2)	Totals
South America	Terra del Fuego/Patagonia	21,200	
	Argentina north of 47.5°S	1385	
	Chile north of 46°S	743	
	Bolivia	566	
	Peru	1780	
	Equador	120	
	Colombia	111	
	Venezuela	3	25,908
North America	Mexico	11	
	USA and Alaska	75,283	
	Canada	200,806	
	Greenland	1,726,400	2,002,500
Africa			10
Europe	Iceland	11,260	
	Svalbard	36,612	
	Scandinavia and Jan Mayen	3174	
	Alps	2909	
	Pyrenees/Mediterranean mountains	12	53,967
Asia and Russia	Commonwealth of Ind. States	77,223	
	Turkey, Iran and Afghanistan	4000	
	Pakistan and India	40,000	
	Nepal and Bhutan	7500	
	China	56,481	
	Indonesia	7	185,211
Australasia	New Zealand	860	860
Antarctica	Subantarctic islands	7000	
	Antarctic continent	13,586,310	13,593,310

Armstrong *et al.*, 1973; Østrem, 1974; Sugden and John, 1976; Benn and Evans, 1998).

4.2.1.1 Ice sheets and ice caps

Large, continental bodies of ice (ice sheets, i.e. Antarctica, Greenland) and small, local bodies of ice (ice caps, i.e. Vatnajökull, Jostedalsbreen) are normally subdivided into *ice domes* (high areas of relatively slow-moving ice), and *ice streams* and *outlet glaciers* (areas with more rapidly flowing ice). An ice sheet covers a large area, and the major part has such a thickness that the subglacial topography is not reflected on the ice-sheet surface. Ice sheets and ice caps are of sufficient thickness to submerge the underlying landscape. Around the margins of ice sheets and ice caps, however, fast-moving ice streams and outlet glaciers are located in valleys and fjords. An area of more than 50,000 km^2 is commonly used for defining ice sheets. Therefore, the ice masses that at present cover Antarctica and Greenland and the former ice masses over Scandinavia, North America, the British Isles, the Barents Sea Shelf and northern Siberia are *ice sheets*, whereas ice masses over Svalbard, Ellesmere Island, Baffin Island and Iceland are designated as *ice caps*.

Box 4.1 Physical properties of ice

When the annual amount of snow and ice exceeds ablation, there is net accumulation; successive layers of snow build up and the deeper layers are transformed into ice as the volume of air-filled pores is reduced and density increases (Paterson, 1994). New-fallen snow has a density of 0.02–$0.2\,g\,cm^{-3}$, while firn (snow that has survived one melt season) has a density of 0.4–$0.83\,g\,cm^{-3}$. Glacier ice has a density of 0.83–$0.91\,g\,cm^{-3}$, while pure ice has a density of $0.917\,g\,cm^{-3}$. The transformation processes, and the time it takes for transformation, depend on climate. Where melting is rare, like in cold polar regions and at high altitudes, the most important factors are wind transport, crystal movement, changes in crystal size and shape, and internal crystal deformation.

The hardness of ice increases with decreasing temperature. At $0°C$ the hardness is 1.5 (Mohr's scale) and increases to 6 at $-70°C$. Ice is considered to belong to the hexagonal system because the crystals are oriented with six more or less regular edges. An ice crystal is built up of parallel layers where the molecules are in hexagonal rings. Glacier ice is polycrystalline, where the crystals lie more or less disorganized with the c-axis pointing in all directions. Glacier ice behaves like a plastic body. The rate of deformation of an ice crystal increases rapidly with the amount of pressure affecting the ice crystal. As a result, a glacier stretches out and thins on steep slopes, and becomes thicker in flatter areas.

4.2.1.2 Ice domes

An ice dome is a nearly symmetrical, broad, upstanding area of an ice sheet or ice cap. Beneath the ice dome, the land surface may either be a topographic high or a depression. At ice-sheet domes, the ice thickness may exceed 3000 m. Beneath ice cap domes, on the other hand, the ice thickness is only several hundred metres. The subglacial topography can be reflected in glacier surface profiles and the location of ice domes.

4.2.1.3 Plateau glaciers

A glacier that covers a rather flat mountain area, and has outlet glaciers into adjacent valleys, may be called a plateau glacier.

4.2.1.4 Outlet glaciers and ice streams

The ice movement in an ice sheet can be divided into *sheet flow* in the central dome areas, and *stream flow* in outlet glaciers and ice streams, characterized by rapidly moving, channelled ice flow. Outlet glaciers draining plateau glaciers normally have the form of valley glaciers. The accumulation area may be difficult to delimit. The longest glacier in the world, the 700 km Lambert Glacier in Antarctica, and the world's fastest moving glacier, Jacobshavn Glacier in western Greenland, are both ice streams. Ice streams can be divided from surrounding ice by the presence of heavily crevassed zones along their margins, for example, on the ice streams of West Antarctica. The flow and morphology of ice streams are the result of conditions at the glacier bed. In particular, the presence of *deformable sediments* and high *porewater pressure* are considered to be important in maintaining rapid ice flow in ice streams. Outlet glaciers and ice streams terminating in the sea are responsible for most of the icebergs and ice-rafted debris reaching the world's oceans. The dynamics of ice streams are also considered to have been important for the stability and break-up of the great mid-latitude ice sheets (Hughes *et al.*, 1985).

4.2.1.5 Icefields

Icefields are different from ice caps because they do not have a dome-like surface and the movement is influenced by the subsurface topography. Icefields are commonly located in areas of gentle but dissected topography.

GLACIER TYPES *51*

The Columbia Ice Field in the Canadian Rocky Mountains, the icefields in the St Elias Mountains in the Canadian Yukon Territory and Alaska, the Kunlun Shan icefields in China, and the Patagonian icefield are all good examples.

4.2.1.6 Valley glaciers

The entire glacier may be located in a valley, or ice masses may discharge from an icefield or cirque into a valley to form a valley glacier. Valley glaciers may be simple, single-branched, or form dendritic networks which can be classified according to their position in their drainage basin. Bedrock slopes are often steep, making the altitudinal ranges of valley glaciers quite large. Snow avalanches from the steep valley sides may increase the accumulation considerably.

4.2.1.7 Transection glaciers

Transection glaciers are interconnected systems of valley glaciers developed in highly dissected landscapes flowing from several directions into a system of radiating valleys. The glaciers may overspill drainage divides and form either diffluenced or confluenced glacier systems.

4.2.1.8 Cirque glaciers

Cirque glaciers are located in armchair-shaped bedrock depressions or hollows. Cirque glaciers may either be entirely confined to the bedrock depressions, or to glaciers that are part of larger valley glaciers in areas with more extensive glacierization. The mass balance on cirque glaciers may be highly influenced by wind-driven snow from adjacent mountain plateaux.

4.2.1.9 Piedmont glaciers

A piedmont glacier flows through a narrow gorge and on to a flat area without any lateral topographical obstacles. The Malaspina Glacier in Alaska and Skeidararjökull in Iceland are both good examples of piedmont glaciers,

characterized by having large areas below the equilibrium line altitude.

4.2.1.10 Ice aprons, ice fringes, glacierets and niche glaciers

Ice aprons are thin accumulations of snow and ice commonly lying on relatively steep mountain sides. Ice fringes are similar ice and snow accumulations occupying small depressions along coasts. Glacierets are referred to as thin patches of ice occupying terrain depressions formed by snow drifting and avalanching. Glacierets have been called fall glaciers or regenerated glaciers where they exist below the ELA as a result of ice avalanches from overlying ice falls. Niche glaciers are controlled by a niche or rock bench in a mountain or valley side. Niche glaciers and ice aprons are different from snow patches because they move as a result of internal deformation and/or basal sliding.

4.2.1.11 Glacier ice shelves

Glacier ice shelves are the floating part of ice sheets. The best examples are located on the margins of the West Antarctic Ice Sheet. In Antarctica, the ice shelves constitute about 7 per cent of the surface area of the ice sheet, 44 per cent of the coastline, and account for approximately 80 per cent of the total ablation. The overall thickness of the edges standing in the sea of the Antarctic ice sheet is approximately 200 m. The Ronne-Filchner and the Ross, the two largest ice shelves, are fed both from the West and East Antarctic ice sheets. The Amery, Ross and Ronne-Filchner ice shelves together drain about 62 per cent of the surface area of the Antarctic continent. The flow rates on these ice shelves are about 0.8–2.6 km yr^{-1}.

4.2.1.12 Sea-ice ice shelves

Sea-ice ice shelves are normally formed by a combination of thickening by snow accumulating on the surface, and freezing of seawater. Movement and deformation in such a shelf is entirely determined by the weight of the ice mass. Sea-ice ice shelves exist where annual temperatures are sufficiently low for ice to form and where embayments and islands

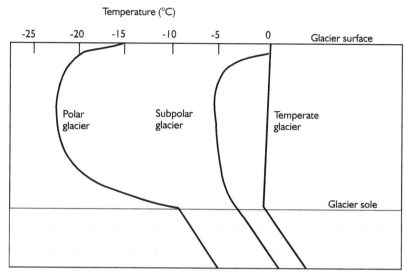

FIGURE 4.1 Tentative temperature profiles through polar, subpolar and temperate glaciers. (Adapted from Skinner and Porter, 1987)

provide anchor points for sea ice to thicken. Examples of sea-ice ice shelves are found in the Canadian and Greenland High Arctic.

4.2.1.13 Composite ice shelves

Composite ice shelves are formed where floating glaciers and sea ice are protected from currents and wave disruption. A good example of such an ice shelf is that at Cape Alfred Ernest on the NW coast of Ellesmere Island in the Canadian Arctic. Composite ice shelves are also present in Antarctica.

4.2.1.14 Ice rises

Ice rises form where floating ice shelves thicken sufficiently to be grounded on offshore shoals and islands. Ice rises are normally formed either by thickening due to surface accumulation or by overriding by the ice shelf. In the first case the ice rise will have a radial flow pattern independent of the general flow direction of the surrounding ice shelf. Where glacier ice shelves override offshore shoals, they normally become heavily crevassed, forming ice rumples. Several examples exist within the Antarctic ice shelves, for example, Roosevelt Island in the Ross ice shelf, and the Gipps Ice Rise in the Larsen ice shelf.

4.2.2 Classification based on physical properties (temperature distribution)

The ice temperature is an important factor for a variety of glacial processes, such as glacier flow, meltwater drainage, and subglacial erosion and deposition. An important distinction is therefore made between *temperate* ice (at the pressure melting point) and *polar* ('cold') ice (below the pressure melting point) (Fig. 4.1). In addition, *subpolar* or *polythermal* glaciers are glaciers which are temperate in their inner and deeper parts, but with cold-based margins (Paterson, 1994). As more information on temperature distribution in glaciers and ice sheets becomes available, it is now recognized that this classification is too simple, and many glaciers are difficult to classify according to this scheme. The terms temperate or *wet-based* ice and cold or *cold-based* ice are now widely used for ice at and below the pressure melting point, respectively.

In polar/cold glaciers, the entire ice mass is below the pressure melting point. The term 'pressure melting point' may be misleading, because the ice may contain 'impurities' (glacier ice consists normally of ice, water, air, salts and carbon dioxide) and therefore there may not be a single distinct melting point determined only by pressure. In subpolar/ polythermal glaciers, the temperature in the ice mass is either below or at the pressure melting point. In temperate glaciers, the major part of the ice mass is at the pressure melting point during the summer season.

There is, however, no clear boundary between the different glacier types. On several glaciers all three temperature conditions may be present simultaneously. Expressions like *dynamically/climatically*, *active/inactive* glaciers as a result of movement and mass exchange are also used. *Maritime* and *continental* glaciers may also be utilized to classify glaciers according to their proximity to the coast and the amount of precipitation falling upon them.

4.3 Temperature distribution in glaciers and ice sheets

Temperatures in glaciers and ice sheets vary in space and time. Temperate glaciers have temperatures at or close to $0°C$, while the upper part of the Antarctic ice sheet may be as cold as -40 to $-60°C$ (Fig. 4.2). The melting temperature of ice decreases with increasing pressure at a rate of $0.072°C$ per 10^6 pascals (MPa; $1\,Pa = 1\,N\,m^{-2}$). As an example, the pressure at the base of a 2000 m thick glacier or ice sheet is about 17.6 MPa, corresponding to a lowering of the melting point to $-1.27°C$. Therefore in glaciers and ice sheets it is more appropriate to use *pressure melting point* rather than melting point. The melting point is also influenced by solutes. The freezing point of normal seawater is approximately $-2°C$ at atmospheric pressure.

The temperature distribution in glaciers and ice sheets is important because it is related to other glaciological and geological processes. The rate of deformation is dependent on temperature. A cooling from -10 to $-25°C$ reduces the deformation rate by a factor of five. Basal ice temperature controls erosion. Glaciers and ice sheets at the pressure melting point are able to erode, but if the bed is frozen to the substratum, glacial erosion may be negligible. If the basal temperature in a

frozen glacier or ice sheet suddenly reaches the pressure melting point, it may start to slide and cause a glacial advance.

The temperature of glacier ice is controlled by three main factors: (a) *heat exchange* with the atmosphere; (b) the *geothermal* heat flux; and (c) *frictional* heat due to ice flow. The surface ice temperature is controlled by the air temperature. Geothermal heat and, if the glacier is sliding, frictional heat influence the basal ice temperature. *Internal deformation* and *refreezing* of meltwater are important factors for the englacial temperature distribution. In addition, conduction, glacier movement and water flow/percolation all transfer heat within the ice mass.

The temperature distribution in glaciers/ice sheets may occur in the following forms:

(a) The entire ice mass is below the melting point.
(b) The ice mass is at the melting point only at the sole.
(c) A basal layer is at melting point while the upper part is below the melting point.
(d) Except for a ca. 15 m surface layer subject to seasonal temperature variations, the ice mass is at the melting point.

Glaciers in category (a) and (b) are termed polar or cold, while category (c) and (d) glaciers are called polythermal and temperate, respectively. Values of thermal parameters for pure ice are listed in Table 4.2.

It has been generally stated that the temperature at a depth of 10–15 m in a glacier or ice sheet is close to the mean annual air temperature. If no melting takes place, temperature changes are transferred into the ice by heat conduction. A firn layer buried by a

TABLE 4.2 Values of thermal parameters of pure ice (Yen, 1981; Paterson, 1994)

Temperature (°C)	0	−50
Specific heat capacity ($J\,kg^{-1}\,K^{-1}$)	2097	1741
Latent heat of fusion ($kJ\,kg^{-1}$)	333.5	—
Thermal conductivity ($W\,m^{-1}\,K^{-1}$)	2.10	2.76
Thermal diffusivity ($10^{-6}\,m^2\,s^{-1}$)	1.09	1.73

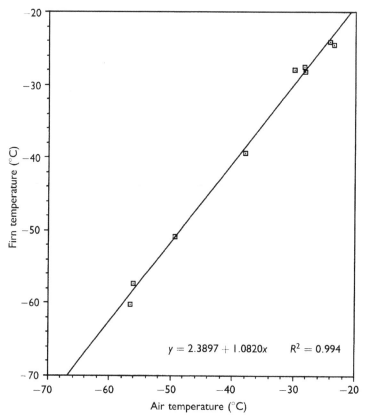

$$y = 2.3897 + 1.0820x \qquad R^2 = 0.994$$

FIGURE 4.2 Regression analysis of annual air temperature and 10 m depth firn temperatures in the dry-snow areas in Greenland and Antarctica (data from Paterson, 1994, Table 10.3: 210)

subsequent year's snowfall will remain at the same temperature and equal to the mean annual temperature at the surface of the ice mass. This is, however, only the case if the maximum air temperature is less than 0°C. Meltwater that refreezes makes the temperature in the firn higher than the mean annual air temperature. The temperature on the glacier surface does not rise above 0°C if the air temperature does. Winter snow reduces heat loss, causing the ice temperature in the ablation zone to be higher than outside the layer of snow from the last winter season. Figure 4.2 shows a regression analysis of annual air temperature and 10 m depth firn temperatures in the dry-snow zone in Greenland and Antarctica (Paterson, 1994).

Since air temperatures commonly decrease with increasing elevation and increasing latitude, so do glacier temperatures. Several factors influence the temperature regime in glaciers, such as the heat that is produced by transformation of mechanical energy within or beneath a glacier. The most important factor, however, is the heat exchange in connection with conduction, radiation and convection. Meltwater transports heat energy to the inner and deeper parts of temperate and subpolar glaciers, causing energy release when this water freezes.

The temperature conditions and distribution in ice sheets are mainly determined by the air temperature in the accumulation zone, the amount of precipitation, glacier dynamics and the geothermal heat flux. Since both temperature and precipitation in addition to glacier dynamics vary significantly between the margins and the central areas, the isotherms (lines through areas with the same temperature) are not parallel with the glacier surface over large areas.

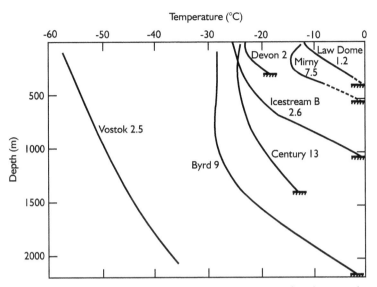

FIGURE 4.3 Measured temperature profiles in accumulation areas of polar ice sheets and ice caps. Broken lines indicate extrapolations. (Modified from Paterson, 1994)

The geothermal heat flux and the frictional heat in the glacier sole may have a significant impact on the glacier movement, since the plasticity is related to temperature. Surging of polar and subpolar glaciers may be explained by these effects. The heat flux from the interior of the Earth is about 50 calories/year/cm^2. This corresponds to melting of an ice thickness of about 7 mm yr^{-1}. In volcanic areas with large geothermal heat fluxes, the heat flow may be a significant factor in subglacial melting.

At the initiation of the melting season, the meltwater in the accumulation area starts to penetrate into the snow, releasing latent heat by freezing. The temperature in the permeable snow pack may therefore ultimately reach 0°C. In temperate glaciers the winter cold may reach down to 5–10 m depth. During the autumn and winter cooling, the water-soaked snow freezes first. This causes a slow downward freezing of the snowpack, and deeper parts of the firn area may therefore remain at the melting point throughout the winter season.

In the ablation area, the situation is different because impermeable ice ensures that the water, except in crevasses and holes, flows off on to the glacier surface. The temperature is therefore mainly determined by conduction.

Because the temperature on the glacier surface does not rise above 0°C, the temperature down in the ice, where annual temperature variations are not detected, is determined by the mean temperature below 0°C on the glacier surface. On temperate glaciers, therefore, ice temperature may be lower in the ablation area than in the accumulation zone.

Figure 4.3 shows measured temperature profiles in glaciers (for references, see Paterson, 1994: 221). Most curves show the features of the theoretical curves based on heat advection. The negative temperature gradients in the upper parts of the Byrd Station and Mirny sites are due to flow of cold ice from greater altitudes. The mean temperature gradient for all the profiles is 1.8°C per 100 m. Temperature gradients at the bed range from 1.8°C per 100 m at Camp Century to 5.2°C per 100 m at Mirny. A geothermal heat flux of 50 mW m^{-2} gives a temperature gradient of 2.4°C per 100 m in the ice (Paterson, 1994).

Since ice sheets never reach steady-state, the temperature distributions will not be in accordance with modelling based on this assumption. Non-steady-state calculations need information concerning, for example, past surface temperature, accumulation rate and ice thickness. Ice cores can now provide pertinent information about these parameters.

Investigations have shown that temperature variations during a glacial cycle could penetrate an ice sheet by means of conduction (Paterson, 1994). As a result, the basal temperature gradient changes with time. Almost all calculations of time-dependent temperatures have been based on the assumption that velocity adjusts to changes in precipitation, and therefore the ice thickness is constant, an assumption which is not true. A doubling of the precipitation rate over the Greenland ice sheet may increase the thickness of the ice by 10–15 per cent over a few thousand years.

4.4 Glacier monitoring

International monitoring of glacier variations started in 1894. At present, the World Glacier Monitoring Service (WGMS) collects standardized glacier information. The data basis includes observations on *specific balance*, *cumulative specific balance*, *accumulation area ratio* (AAR), the *equilibrium line altitude* (ELA), and changes in length. Most of the data and the longest records come from the Alps and Scandinavia.

About 90 years of observations reveal a general shrinkage of mountain glaciers on a global scale after the Little Ice Age (ca. AD 1650–1930). This was most pronounced during the first half of the twentieth century. More recently, however, glaciers have begun to grow in several regions, for example, in maritime western Scandinavia and in New Zealand. Important empirical information has started to become available on the complex relationship between climate and glaciers (e.g. Haeberli *et al.*, 1989; Oerlemans, 1989).

Records of glacier fluctuations compiled by the WGMS have been used by Oerlemans (1994) to derive an independent estimate of global warming during the last 100 years. The retreat of glaciers during the last 100 years appears, with a few exceptions, to be coherent over the globe. Modelling of the climate sensitivity of glaciers reveals that the observed glacier retreat can be explained by a linear warming trend of 0.66 Kelvin per century (Oerlemans, 1994).

During the last 25 years, several remote sensing techniques have been applied to the study of glaciers. These techniques include measurements of ice thickness by *radio-echo sounding* from surface and airborne platforms, changes in surface elevation over time with *aerial photogrammetric methods* and by *geodetic airborne* and *spaceborne radar* and *laser altimetry*, declination of the surface expression of glacier facies with satellite sensors, and measurements of the fluctuations in the fronts of valley glaciers and outlet glacier margins at ice fields, ice caps and ice sheets.

Airborne scanning laser altimetry is a relatively new technique for remote sensing of ground elevation (e.g. Kennett and Eiken, 1997). A laser ranger scans across a swath beneath the aircraft, giving a 2D distribution of altitude when combined with data on position and orientation of the aircraft. Smooth snow-covered glaciers are the best for laser scanning altimetry since they are highly reflective. Results from Hardangerjøkulen, central southern Norway (Kennett and Eiken, 1997) show that noise levels are very low (ca. 2 cm). Overlapping swaths show repeatability of ±10 cm. The high accuracy and coverage (about 20,000 points per km^2) enables reliable measurement of glacier volume changes. Scanning laser altimetry has many advantages compared with photogrammetry.

4.5 Glacier monitoring by satellites

The *Earth Observing System* (EOS) has been developed to build a uniform database covering most glaciers of the world and to monitor changes in glaciers on a periodic basis (Kieffer *et al.*, 1994). This planned survey will build on current and previous projects, including 20 years of space-based observations of glaciers, such as observations by LANDSAT, SPOT, ERS and RADARSAT. The satellite can observe all of the Earth's surface at latitudes less than 85°. The project is presently developed to acquire annual images of most of the world's glaciers to monitor, for example, surface velocities, areal extent and the position of large crevasses, to a precision of around 5 m. This project will hopefully result in a uniform, global database at 15 m resolution. This collaborative effort could potentially accomplish a periodic survey of the world's glaciers resulting in a uniform image data-set, monitor special events such as glacier surges, produce maps of the areal extent of glaciers and snow fields, give information about glacier surface velocities, advance and retreat, and give an inventory of the glaciers, including mean surface speed, length, width, areal extent, snowline, and the temporal changes in these parameters (Kieffer *et al.*, 1994).

4.6 Determination of the equilibrium line altitude (ELA)

The equilibrium line altitude marks the area or zone on the glacier where accumulation is balanced by ablation. The ELA is sensitive to variations in winter precipitation, summer temperature, and wind transport of dry snow. When the annual net mass balance is negative, the ELA rises, and when the annual net balance is positive, the ELA drops. The steady-state ELA is defined as the ELA when the annual net balance is zero, and can be calculated by linear regression analysis of annual net balance data and corresponding ELAs over some years (Fig. 4.4).

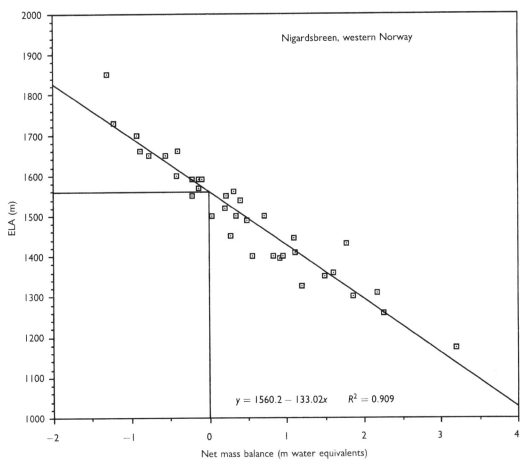

FIGURE 4.4 The steady-state ELA (= 1560 m) at Nigardsbreen, an eastern outlet glacier from Jostedals-breen, western Norway. Based on data (observation period 1962–1997) in Kjøllmoen (1998)

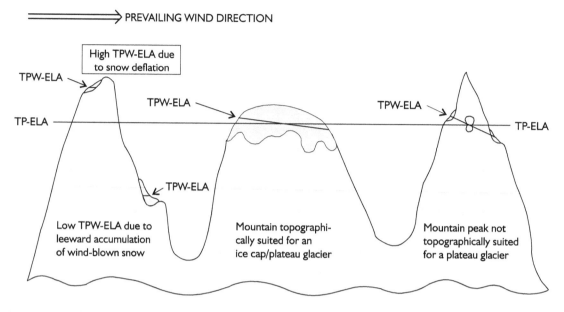

Figure 4.5 Schematic examples showing the difference between the TP-ELA (temperature–precipitation equilibrium line altitude) at plateau glaciers and the TPW-ELA (temperature–precipitation–wind equilibrium line altitude) at cirque glaciers. (Modified from Dahl and Nesje, 1992)

The climatic ELA is the average ELA over a 30-year period (corresponding to a 'normal' climatic period).

Climatic processes influencing the ELA on glaciers commonly involve ablation (mainly determined by the summer temperature) and the winter accumulation (reflecting the winter precipitation) giving the TP-ELA (see below). In addition, however, wind transport of dry snow is an important factor for the glacier mass balance. On plateau glaciers, snow deflation and drifting dominate on the windward side, while snow accumulates on the leeward side. By calculating the mean ELA in all glacier quadrants, the influence of wind on plateau glaciers can be neglected. The resulting ELA is therefore defined as the TP-ELA (temperature/precipitation ELA). The TP-ELA reflects the combined influence of the regional ablation-season temperature and accumulation-season precipitation (Dahl and Nesje, 1992; Dahl *et al.*, 1997).

In deeply incised cirques and valleys surrounded by wide, wind-exposed mountain plateaux, the snow may deflate from the plateaux and accumulate in the cirques and valleys, either by direct accumulation on the cirque/valley glaciers, or by avalanching from the mountain slopes. This may thereby increase significantly the accumulation on the cirque/valley glaciers (Dahl and Nesje, 1992; Tvede and Laumann, 1997). Consequently, the mean ELA on a plateau glacier (average for all quadrants) defines the TP-ELA, while the ELA on a cirque glacier, commonly influenced by wind-transported snow, gives the TPW-ELA. Therefore, the TPW-ELA is commonly lower than the TP-ELA (Fig. 4.5).

4.7 Reconstruction of the equilibrium line altitude

Fluctuations in the ELA provide an important indicator of glacier response to climate change which may allow reconstructions of palaeoclimate, but also of future glacier response to given climate change.

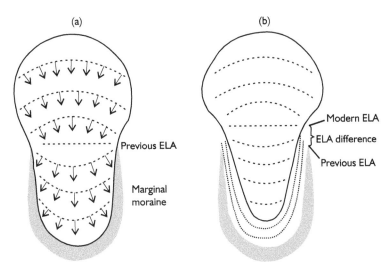

FIGURE 4.6 The principle of calculating the depression of the equilibrium line altitude on a glacier based on the maximum elevation of lateral moraines. The previous extent (a) is compared with the modern extent (b) for an idealized glacier. Dashed lines indicate surface contours and arrows indicate ice flow direction. (Modified from Nesje, 1992)

The most common approaches in reconstructing palaeo-ELAs are to use:

(a) the maximum elevation of lateral moraines (MELM);
(b) the median elevation of glaciers (MEG);
(c) the toe-to-headwall altitude ratio (THAR);
(d) the ratio of the accumulation area to the total area (AAR); and
(e) the balance ratio method.

These will be considered in turn.

(a) *Maximum elevation of lateral moraines.* Due to the nature of glacier flow towards the centre and the margin of the glacier above and below the ELA, respectively, lateral moraines are theoretically only deposited in the ablation zone below the ELA. As a result, the maximum elevation of lateral moraines reflects the position of the corresponding ELA (Fig. 4.6).

Commonly, however, it is difficult to assess whether or not a lateral moraine is preserved entirely in the upper part or whether moraine deposition started immediately down-glacier of the ELA. Consequently, ELA estimates derived from eroded and/or non-deposited lateral moraines may be too low. In contrast,

the assumption that the maximum altitude of lateral moraines is obtained during steady-state conditions can overestimate the ELA. If initial glacier retreat is slow, additional moraine material could be deposited in the prolongation of the former steady-state lateral moraine. A continuous supply of debris from the valley or cirque walls may lead to the same source of error in the ELA calculations.

(b) *Median elevation of glaciers.* The median elevation of glaciers (MEG) has been used to estimate ELAs. However, empirical evidence from modern glaciers suggests that the MEG overestimates the ELA. In addition, this method fails to take into account variations in valley morphology, which strongly affect the area–elevation distribution of a glacier. However, it works well for small glaciers with even area/altitude distributions. Still, the main problem is to define the headward limit of a former glacier.

(c) *Toe-to-headwall altitude ratio.* This ratio between the maximum and minimum altitude of a glacier has been used as a quick estimate to calculate the ELA. Ratios of 0.35–0.4 normally

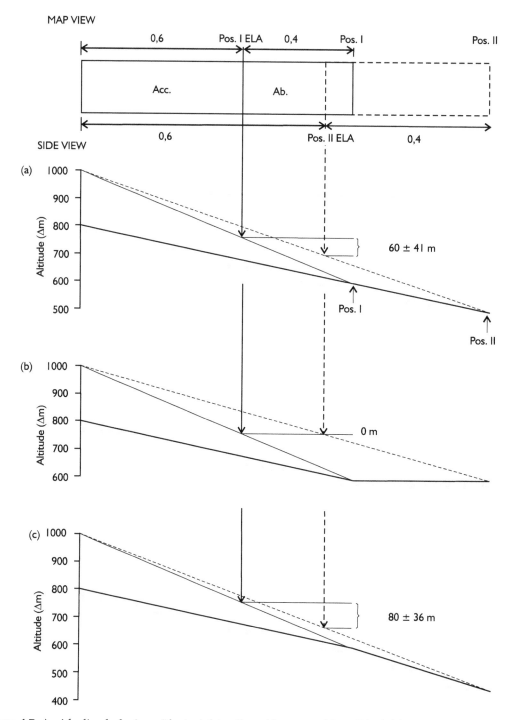

FIGURE 4.7 An idealized glacier with straight valley sides at positions I (solid line) and II (punctuated line). The ELA difference between positions I and II, using the AAR approach at different slope angles, is shown. The figure illustrates the importance of surface slope angles of the underlying topography when applying the AAR approach for calculating former ELAs. (Adapted from Nesje, 1992)

give the most correct estimates. Again, a major problem is to define the headward limit of a former glacier.

(d) The *ratio of the accumulation area to the total area* (accumulation area ratio, AAR) is based on the assumption that the steady-state AAR of former glaciers is 0.6 ± 0.05 (Porter, 1975), a value derived from temperate glaciers from different regions of the world, mostly NW North America.

The AAR of a glacier varies mainly as a function of its mass balance; ratios below 0.5 indicate negative mass balance, 0.5–0.8 correspond to steady-state conditions, and values above 0.8 reflect positive mass balance regimes (Andrews, 1975). An AAR of 0.6 ± 0.05 is generally considered to characterize steady-state conditions of valley/cirque glaciers. Ice caps and piedmont glaciers may, however, differ significantly from this ratio. The largest source of inaccuracy related to the AAR method of determining the ELA on former glaciers is the reconstruction of the surface contours, especially if the glacier margins intersect valley-side topographic contours at small angles or coincide with them for some distance. However, this source of error is considered to be randomly distributed and is not considered to introduce major deviations from representative conditions. In addition, this method only requires glacier reconstruction as high as the former ELA. A theoretical evaluation of the AAR approach, using changing slope angles and valley morphology on idealized glaciers, shows that glaciers advancing into flat areas underestimate the ELA depression, while glaciers moving into areas of increasing slope angle overestimate the climatic ELA difference (Fig. 4.7) (Nesje, 1992). Consequently, topographical and morphological effects on calculated ELA depressions on glaciers must be carefully evaluated.

(e) *Balance ratio method*. As demonstrated above, one shortcoming of the AAR method, and also the MEG approach, is that they do not fully account for variations in *hypsometry* (distribution of glacier area over its altitudinal range). To overcome this problem, a balance ratio method was developed by Furbish and Andrews (1984). This approach takes account of both glacier hypsometry and the shape of the mass balance curve and is based on the fact that, for glaciers in equilibrium, the total annual accumulation above the ELA must balance the total annual ablation below the ELA. This can be expressed as the areas above and below the ELA multiplied by the average accumulation and ablation, respectively (for further details, see Furbish and Andrews, 1984; Benn and Evans, 1998: 84).

4.8 Mass balance

Glaciers and ice sheets are stores of water, exchanging mass with other components involved in the global hydrological system. Glaciers and ice sheets grow by snow and ice accumulation, and lose mass by different ablation processes. The difference between accumulation and ablation over a given time span is the mass balance, which can be either positive or negative. The mass balance reflects the climate of the region, together with glacier morphology and local topographic conditions. Mass balance measurements can therefore give information on the causes of retreat or advance of glaciers.

One of the first systematic analyses of the annual mass budget of a glacier was made by Ahlmann (1927). Statistical relationships between mass balance and meteorological parameters have been investigated on several glaciers (Letréguilly, 1988; Pelto, 1988), and the physical relationships studied by Holmgren (1971), Kuhn (1979) and Braithwaite (1995), amongst others.

Mass-balance variations can be associated with atmospheric circulation, linking them to atmospheric changes rather than single meteorological parameters. This approach was used by Hoinkes (1968) to show how glacier variations in Switzerland were related to cyclonic and anticyclonic conditions. Alt (1987) found that extreme mass balance years at the Queen Elisabeth Island ice caps, Canada, were related to the position of the Arctic front. In southwestern Canada, Yarnal (1984) found that two glaciers were sensitive to both large- and small-scale synoptic weather situations. Voloshina (1988) discussed why the position of the Siberian anticyclone forms an inverse relationship between the mass balance for glaciers in northern Scandinavia and in the northern Urals. The strength of the Aleutian low is important for the determination of the storm track and high mass balance in the Alaskan Range and the Cascades (Walters and Meier, 1989). McCabe and Fountain (1995) found that the winter balance of the South Cascade Glacier correlates to the pressure difference between the Gulf of Alaska and the west coast of Canada. Finally, Pohjola and Rogers (1997a,b) used atmospheric circulation and synoptic weather studies to explain variations in glacier mass balance on Scandinavian glaciers. They also demonstrated that a high net balance on Storglaciären, the glacier with the longest mass-balance record in the world, is favoured by strong westerly maritime air flow which increases the winter accumulation. Holmlund and Schneider (1997) used a continentality index as a measure of the nature of climate, mass balance, and glacier-front response along a west–east transect in a region just north of the Arctic Circle in Scandinavia. These studies demonstrate the potential of the relationship between glacier mass balance and synoptic weather studies. This is important when using glacier-front or ice-core records to reconstruct past atmospheric circulation.

The most important accumulation factor on glaciers is snowfall. The amount and distribution may, however, vary considerably geographically and seasonally. The highest accumulation rates are observed in maritime, mountainous regions with frequent winds blowing in from the sea, for example, in western North America, the west coast of New Zealand, western Patagonia, southern Iceland and western Scandinavia. In contrast, snowfall is lowest far away from oceanic sources and in precipitation 'shadows' in downwind positions relative to high mountains. Locally, accumulation may be strongly influenced by wind transport of dry snow and by snow avalanches. Ice and snow crystals, or rime ice, can also form on glacier surfaces by freezing of wind-transported, supercooled vapour or water droplets. This process is most common on maritime glaciers.

The glacier accumulation zone has been divided according to melting and refreezing (Fig. 4.8). The dry snow zone is below $0°C$ and therefore no meltwater is present. The dry snowline separates the dry snow zone from the percolation zone. The percolation zone is characterized by some surface melting, and the water percolates through the snow where it refreezes. The percolation depth normally increases with decreasing altitude. The wet snowline marks the upper boundary of the wet snow zone, where the snow temperature is $0°C$.

In some places, most commonly in the lower areas, the refrozen meltwater may form a continuous layer of superimposed ice, called the *superimposed ice zone*. The equilibrium line marks the zone where the annual accumulation at the end of the ablation zone is balanced by the total ablation.

Ablation refers to the processes causing mass loss from the glacier, including wind deflation, avalanching from the front, calving of icebergs, from runoff melting, evaporation and sublimation. Wind deflation is wind-scouring of snow resulting in the removal of snow and ice from the glacier surface. The process is most efficient in areas of strong katabatic winds and on narrow valley glaciers. Avalanching may be an important ablation factor, especially where the ice front terminates above steep rock cliffs. Ice that breaks off from the glacier front falls down, and if the avalanching rate is greater than the melting rate, regenerated glaciers may form below. Iceberg calving is the mass loss at the margins of glaciers and ice sheets terminating in water (lake or sea). Calving events may vary considerably in scale, from small blocks to enormous icebergs. In March 1990, for example, a 3.5 km long iceberg, weighing approximately 100 million tonnes, of the Erebus Glacier in Antarctica broke off. In January 1995, a portion of the Larsen ice shelf in the Antarctic Peninsula broke up, causing a marginal retreat of 2 km in five days. On a global scale, calving is an important ablation process since large portions of the Antarctic ice sheet terminate in the sea.

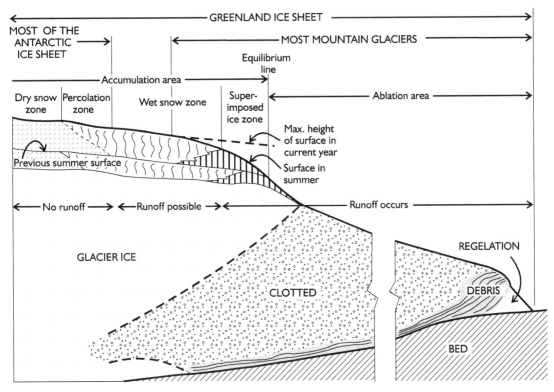

FIGURE 4.8 Subdivision of glacier accumulation zones according to patterns of melting and refreezing. (Adapted from Menzies, 1995)

Melting, evaporation and sublimation are processes causing transformation of ice to water, water to vapour, and ice to vapour, respectively. These processes take place if there is extra energy available at the glacier surface when the temperature has been raised to the melting point. A net deficit of energy, on the other hand, can lower the ice temperature or cause ice accumulation by condensation of vapour or freezing. The energy balance is the surplus or deficit of energy over time, and is an important factor for ablation rates (Paterson, 1994). Energy balance factors on a glacier surface are solar radiation, long-wave radiation, sensible and latent heat, freezing, condensation, evaporation and sublimation.

Solar radiation reaches the surface as direct sunshine or as diffuse radiation scattered through the atmosphere. Some of the radiation is reflected, and the percentage that is reflected from the surface is termed *albedo*. The albedo is high for new-fallen snow and low for dirty glacier surfaces (Table 4.3).

The *short-wave radiation* is dependent on its aspect. Radiation is highest when the sun's rays make an oblique angle with the surface. The low solar angle in the mid- and high latitudes during winter reduces incidence

TABLE 4.3 Albedo values (in per cent) for different types of snow and ice (from Paterson, 1994)

	Range	Mean
Dry snow	80–97	84
Melting snow	66–88	74
Firn	43–69	53
Clean ice	34–51	40
Slightly dirty ice	26–33	29
Dirty ice	15–25	21
Debris-covered ice	10–15	12

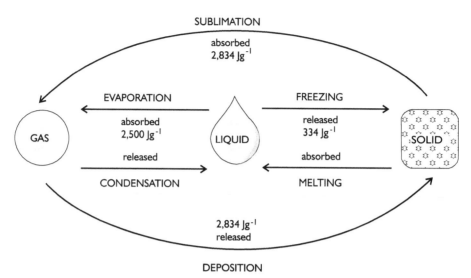

FIGURE 4.9 Phase changes between ice, water and vapour. The amount of latent heat energy consumed and released by the transformation is shown. (Modified from Benn and Evans, 1998)

compared with the tropics. Locally the solar receipt pattern is modified by surface gradient, aspect, and mountain shading.

Long-wave energy is emitted from the atmosphere, bedrock surfaces and other heated surfaces. Long-wave radiation is an important energy budget component when the air is humid. Dry, clear air has a lower ability to trap long-wave radiation. Along the margins of valley glaciers, where radiation is emitted from dark rock surfaces, long-wave radiation may be an important ablation process.

Thermal energy exchanged at the interface between the atmosphere and the glacier surface is termed *sensible heat* transported by warm air masses, such as valley winds or föhn-winds on the lee sides of mountains, or winds accompanying cyclones. Transfer of sensible heat is most efficient when the air is much warmer than the ice and snow surfaces, and where strong, turbulent winds blow over a rough glacier surface (Paterson, 1994).

Changes between ice, water and vapour consume energy in the form of latent heat. Melting consumes 334 joules per gram of ice melted. Evaporation consumes over eight times as much ($2500 \, \text{J} \, \text{g}^{-1}$). Freezing and condensation release the same amount of energy (Fig. 4.9). Freezing of rainwater or condensation of

water vapour on a glacier surface can transfer considerable amounts of energy.

The relative importance of each of the energy balance components varies both temporally and spatially. Commonly, the net radiation (both short- and long-wave) is the most important component, the highest proportions being associated with clear skies. In areas with a continental climate, net radiation has been calculated to amount to more than 60 per cent of the ablation energy. In more humid, maritime climates, this value may be reduced to 10–50 per cent.

Debris on the surface of snow and glacier ice influence ablation rates in two ways. Rock surfaces can heat up and re-emit long-wave radiation, causing melting of adjacent ice and snow. If the debris layer, on the other hand, is thicker than 1–2 cm, the debris will protect the ice and snow from ablation. On glaciers with a thick debris cover, the ablation may be negligible.

The amount of snow and ice stored in glaciers is subject to systematic changes during a year, due to cycles of accumulation and ablation. Several types of cycles occur, depending on the timing of warm and cold seasons, maximum precipitation, and variations in the proportion of precipitation falling

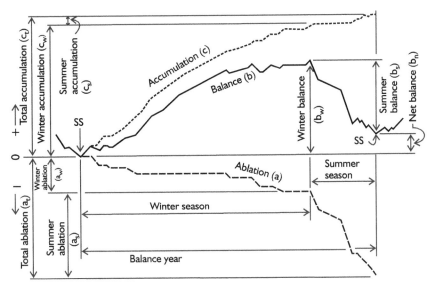

Figure 4.10 Terms used in mass-balance studies for one balance year, adapted from UNESCO (1970). See Glossary of terms used in mass balance studies in Box 4.2

as snow. The most common cycles are: (a) winter accumulation type, with a well-defined winter accumulation season and summer ablation season; (b) summer accumulation type, with maxima in accumulation and ablation taking place at the same time during the summer season; and (c) year-around ablation type, with one or more accumulation maxima coinciding with wet seasons.

The mass balance of a glacier is measured at representative points on its surface. The results of the mass balance measurements are integrated and reported as a value averaged for the whole glacier surface, so that comparisons may be made between different glaciers. The mass balance components are expressed in metres of water equivalents.

The methodologies and techniques used to measure glacier mass balance commonly follow guidelines from the *Commission on Snow and Ice of the International Association of Scientific Hydrology* (UNESCO, 1970). The different terms used are illustrated in Fig. 4.10.

The winter balance is commonly measured in April and May by sounding the snow depth at several points on the glacier surface. The soundings always refer to the last summer surface, which may consist either of glacier ice or firn, depending on where you are on the glacier. The density of the snow is measured at a few sites, preferably at different elevations. The water equivalents are thereafter calculated on the glacier. The points are plotted on a map and isolines of winter accumulation are drawn. Usually, some snow falls on the glacier after the measurements of the winter accumulation are finished. This additional accumulation may be measured, but the most common approach is to calculate it from precipitation and temperature measurements at meteorological stations close to the glacier.

The summer balance is calculated at several stakes drilled into the glacier surface by measuring the lowering of the snow/ice surface during the ablation season. The summer balance measured at the stakes is then transferred to a glacier map, and isolines of the summer balance can be drawn. The summer balance is commonly more evenly distributed than the winter balance, since in most cases it decreases with rising elevation. The net balance is calculated as the winter balance minus the summer balance ($b_n = b_w - b_s$).

Box 4.2 Glossary of terms used in mass balance studies

Ablation: all processes that reduce the glacier mass, including calving.

Ablation zone: the part of the glacier where summer melt exceeds winter accumulation. Not only does this include the total melting of the snow cover of the last winter, but also a layer of glacier ice. A deficit of mass appears in that area. The zone lies at lower altitudes of the glacier surface. The ablation zone meets the accumulation zone at the equilibrium line.

Accumulation: all processes that increase the glacier mass. Winter snowfalls are the most important source of mass gain. Redeposition of snow by wind and avalanche are important factors on cirque glaciers and on glaciers surrounded by large mountain plateaux and steep valley sides.

Accumulation area ratio (AAR): the ratio of the accumulation zone to the entire glacier with respect to any particular year. The AAR is an indicator of the glacier balance state in the observation year. The ratio is expressed as a proportion of the total area of the glacier.

Accumulation zone: the part of the glacier where snow that has accumulated during the winter does not melt completely in the subsequent summer. An increase of mass is observed in this area. The zone lies normally in the upper part of the glacier. The accumulation zone meets the ablation zone at the equilibrium line.

Annual ablation: the mass loss during one measurement year in the fixed date system.

Annual accumulation: the mass gain to the glacier during one measurement year in the fixed date system.

Annual balance: the sum of the annual accumulation (positive) and the annual ablation (negative) at the end of the measurement year (balance year). This term is used in the fixed date system for measuring and reporting mass balance. Total values are averaged over the entire glacier surface and presented in terms of equivalent water layer (in metres).

Balance year: the time between dates of formation of two consecutive summer surfaces, commonly understood as the time between the beginning of the winter accumulation and the end of the ablation in the subsequent summer (the date of the minimum summer balance). The balance year is rarely exactly equal to one calendar year.

Calving: the process of mass loss in respect of tidewater glaciers (glaciers terminating in the sea or in a lake) and ice shelves (detachment of icebergs).

Climatic equilibrium line: the mean annual equilibrium line over a 30-year period.

Combined system: system of mass balance studies based on a combination of the fixed date system, stratigraphic systems and other direct data to obtain a measure of glacier summer balance, winter balance and net balance.

Cumulative mass balance: the mass balance summed from particular years of an observation period, which indicate the tendency of a glacier mass to have either grown or shrunk.

Equilibrium line: a line joining points on a glacier surface where winter balance equals summer balance. Normally this is a line or narrow zone where the summer melting entirely removes the winter snow cover but not any older ice or firn below this. The line separates the accumulation zone from the ablation zone.

Equilibrium line altitude (ELA): the altitude at which the equilibrium line is noted at the end of any particular balance year. Normally, it is an averaged value with respect to the whole glacier. ELA is used as an indicator of the glacier mass balance state; when it is higher, the net balance is lower and *vice versa*.

Firn: old, coarse-grained snow that has survived at least one summer melt season.

Fixed date system: a system of mass balance study, based on field measurements on the same date in consecutive years.

Glaciation threshold: the critical level where a glacier can form and is normally calculated by means of the 'summit method' (between the lowest mountain carrying a glacier and the highest mountain without a glacier).

Internal accumulation: the water melted out at times of ablation usually drains from the glacier and its mass is thereby reduced. However, in areas with snow or firn temperatures below zero, meltwater percolating through the summer surface can refreeze and thereby add mass to the lower layers of snow or firn.

Mass balance: the change in mass at any point on a glacier surface at any time (positive or negative mass balance). Normally it means a change in the mass of the entire glacier in a standard unit of time (the balance year or measurement year).

Measurement year: the unit of time used in the fixed date system of mass balance study, which is usually taken at the end of the summer or the beginning of winter and lasts 365 days.

Net balance (b_n): the sum of the winter balance (positive) and the summer balance (negative) through the balance year ($b_n = b_w + b_s$). The term is used in the stratigraphic system of measurement and reporting of mass balance. Total values are averaged over the entire glacier area and presented in terms of an equivalent water layer (thickness in metres).

Steady-state equilibrium line: the equilibrium line altitude (ELA) where the net balance (b_n) is zero.

Stratigraphic system: a system of mass balance study based on recognition of the glacier summer surface and the maximum values of accumulation (winter balance) and ablation (summer balance) during the balance year.

Summer balance (b_s): the change in mass (commonly negative) during the summer season. It is usually measured at the end of the summer season and at the time of formation of the summer surface (minimum balance). The term is often used synonymously with summer ablation.

Summer surface: the glacier surface formed as a result of the summer balance. This represents the surface of the minimum glacier volume during the balance year.

Temporary equilibrium line altitude: the equilibrium line at an arbitrarily chosen time of the year. During early spring the temporary ELA is in the lower glacier area, while it is higher up the glacier later in the ablation season.

Winter balance (b_w): the maximum balance value (positive) during the balance year in the stratigraphic system (considered synonymous with winter accumulation). The time when the maximum balance is measured divides the balance year into the winter and summer seasons.

The net balance is positive if the winter balance is greater than the summer balance, and negative if the summer balance is greater than the winter balance. The ELA is the zone on the glacier where the net balance is zero.

Figure 4.11 shows the exponential relationship between mean ablation-season temperature *t* (1 May–30 September) and winter accumulation *A* (1 October–30 April) at the ELA of modern Norwegian glaciers (Liestøl in Sissons, 1979a; Sutherland, 1984), and expressed by the regression equation (Ballantyne, 1989):

$$A = 0.915 \, e^{0.339t} \qquad (r^2 = 0.989, \; P < 0.0001)$$

$$(4.1)$$

where *A* is in metres water equivalent and *t* is in °C. The positive correlation between these two variables for different glaciers reflects the fact that higher levels of mass turnover at the

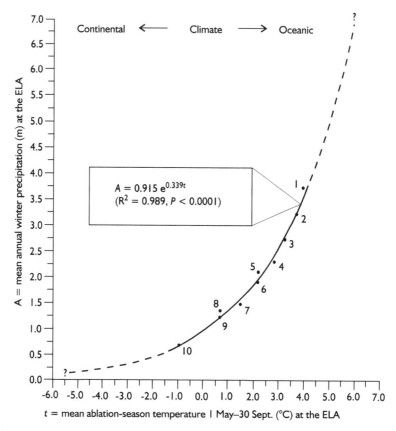

FIGURE 4.11 Mean summer temperatures plotted against accumulation (in metres water equivalent) at the equilibrium line for ten Norwegian glaciers (1, Ålfotbreen; 2, Engabreen; 3, Folgefonna; 4, Nigardsbreen; 5, Tunsbergdalsbreen; 6, Hardangerjøkulen; 7, Storbreen; 8, Austre Memurubreen; 9, Heillstugubreen; 10, Gråsubreen). (Modified from Sutherland, 1984; Dahl *et al.*, 1997)

ELA require higher ablation and thus higher summer temperatures to balance the annual mass budget. This relationship, which is of global application, has also been demonstrated by Loewe (1971) and Ohmura *et al.* (1992). The scattering of the data points in these compilations are due to the fact that they include glaciers where non-climatic factors heavily influence the mass balance.

A similar approach was used to expand the range of summer temperature and winter precipitation of this glacier/climate relationship by using annual winter (1 October–30 April) accumulation measurements and summer (1 May–30 October) temperature at the ELA in the corresponding years calculated from adjacent meteorological stations. The four glaciers used were Ålfotbreen, Hardangerjøkulen, Hellstugubreen (all three in southern Norway), and Brøggerbreen (Svalbard) together with summer temperature data from the adjacent meteorological stations Sandane, Finse, Øvre Tessa, and Isfjord Radio, respectively (Fig. 4.12).

Mass balance data from 14 glaciers in different climatic regimes worldwide (Table 4.4, p. 71) were used to test whether there is a relationship between net mass balance variations and ELA variations. Regression analyses show that there is a fairly good correlation ($R^2 = 0.80$) between these two parameters. An ELA depression of 100 m, taken as a typical Little Ice Age value, indicates a net mass balance increase of about 20 m water equivalents

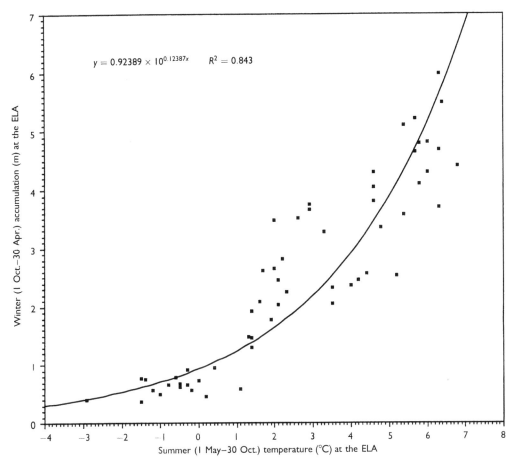

Figure 4.12 Annual winter (1 October–30 April) accumulation measurements and summer (1 May–30 October) temperature at the ELA in corresponding years calculated from adjacent meteorological stations. The ELAs at the four glaciers used were Ålfotbreen, Hardangerjøkulen, Hellstugubreen (all three in southern Norway), and Brøggerbreen (Svalbard) were used, together with summer temperature data from the adjacent meteorological stations Sandane, Finse, Øvre Tessa and Isfjord Radio, respectively

(Fig. 4.13, p. 72, top panel). A depression of the ELA of 400 m (a typical value for the Younger Dryas ELA depression in western Norway) indicates a cumulative net mass balance of approximately 70 m water equivalents (Fig. 4.13, middle panel). A depression of the ELA of 1000 m (suggested Late-glacial maximum ELA depression) indicates, according to this relationship, a net balance increase of about 170 m water equivalents (Fig. 4.13, bottom panel).

Glacier mass balance can also be calculated by measuring other parameters, such as precipitation and runoff. Thus, the net balance

(b_n) of a glacier can be expressed as:

$$b_n = P - R - E \qquad (4.2)$$

where P is precipitation, R is runoff and E is evaporation. This approach to calculating glacier mass balance is termed the hydrological method.

Where detailed mass balance data are not available, a statistical approach can be adopted to estimate ablation rates using mean annual or monthly temperatures or *positive degree days*, defined as the sum of the mean daily temperature for all days with temperatures above 0°C.

Box 4.3 How to calculate winter precipitation from the equilibrium line altitude on glaciers when summer temperature is known

Based on the regression equation 4.1, mean winter precipitation (*A*) can be quantified when mean ablation-season temperature (*t*) is known (see Dahl and Nesje, 1996, for further details). The procedure calculates what mean winter precipitation is or has been at the present ELA of a glacier in steady state. Variations in winter precipitation at other elevations can be calculated by using a precipitation gradient of 8 per cent per 100 m (Haakensen, 1989; Dahl and Nesje, 1992; Laumann and Reeh, 1993). As a first example, if we want to quantify the present mean winter precipitation at an ELA of 1640 m on a glacier, we apply the following procedure. Temperature is lowered by an adiabatic lapse rate of 0.6°C/100 m. If we use a climatic station at an altitude of 1224 m with a mean ablation-season temperature of 4.35°C, the present mean ablation season temperature (1961–90) at the ELA is 1.85°C. Substitution in equation 4.1 gives the following expression:

$$A = 0.915\,e^{(0.339 \times 1.85)}$$

$$= 1.71312, \text{ or ca. } 1.71\,\text{m}$$

As a second approach we wish to quantify mean winter precipitation. For the time period we want to calculate, the mean ablation-season temperature was 1.35°C warmer (remember to adjust for isostatic movements if appropriate), while the contemporaneous ELA was 60 m lower (corresponding to 0.35°C) than at present. The mean ablation-season temperature at the ELA during the time interval in question is thus calculated to be 3.55°C (present mean ablation-season temperature at the ELA of 1.85°C + warmer mean ablation-season temperature during the specific time interval of 1.35°C + warmer mean ablation-season temperature due to a lower ELA of 0.35°C). Put into equation 4.1, this yields the following expression:

$$A = 0.915\,e^{(0.339 \times 3.55)}$$

$$= 3.0484 \text{ or ca. } 3.05\,\text{m}$$

If the present mean winter precipitation of ca. 1710 mm corresponds to 100 per cent, this indicates a mean winter precipitation of approximately 175 per cent during the specific time interval used in this example.

For some Norwegian glaciers, Laumann and Reeh (1993) found melt rates of 3.5–5.6 mm of water per positive degree day for snow, and 5.5–7.5 mm of water per positive degree day for ice. The difference is due to the higher albedo of snow. Melt rates per degree day are higher for maritime glaciers, because higher wind speeds and humidity cause more melting due to transfer of sensible heat and the latent heat of condensation.

Glacier mass balance can also be calculated from aerial photographs and satellite images obtained from successive years or over longer periods. Changes in glacier volume can be measured by changes in the altitude of the glacier surface. This can be converted into mass of water by estimating or measuring the density of snow, firn and ice on different parts of the glacier. High-quality aerial photographs and satellite images are quite expensive to obtain, but they make it possible to study mass balance variations in very remote areas. So far, the radar altimetry used for studies of altitudinal variations of glacier surfaces has not been accurate enough for precise estimates. However, the use of laser altimetry gains sufficient precision for such investigations.

On most glaciers, the amounts of annual ablation and accumulation vary quite systematically with altitude. The rates of which annual accumulation and ablation change with altitude are termed the accumulation and ablation

TABLE 4.4 Mass balance data from 14 glaciers in different climatic regimes

Glacier	Cum b_n (m)	Steady-state ELA (m)	Mean ELA (m)	Difference (m)
1. Place	−19.601	2080	2231	+151
2. White	−4.341	930	1018	+88
3. Ålfotbreen	+14.90	1180	1132	−48
4. Nigardsbreen	+16.96	1560	1494	−66
5. Hardangerjøkulen	+7.54	1680	1595	−85
6. Gråsubreen	−7.41	2060	2121	−61
7. Austre Brøggerbreen	−12.58	265	403	−138
8. Midtre Lovénbreen	−9.98	290	400	+110
9. Obruchev	−1.67	525	528	+3
10. Maliy Aktru	−1.24	3140	3149	+9
11. Ts Tuyuksyskiy	−14.348	3740	3814	+74
12. Urumqihe S. No.1	−5.196	4030	4047	+17
13. Hintereisferner	−17.466	2920	3002	+82
14. Silvretta	−0.163	2760	2765	+5

gradient, respectively. Together, they are defined as the mass balance gradient. Mass balance gradients for some North American glaciers are shown in Fig. 4.14 (p. 73). Steep mass balance gradients are the result of heavy snowfall in the accumulation area and high ablation rates near the front, characteristic of maritime glaciers. Low mass balance gradients, on the other hand, indicate small differences in mass balance with altitude, characteristic of slow-moving, low-gradient, continental glaciers.

On valley and cirque glaciers, the net annual accumulation usually increases with increasing altitude. In western Norway, the *precipitation elevation gradient* is in general ca. 8 per cent per 100 m (Haakensen, 1989; Dahl and Nesje, 1992; Laumann and Reeh, 1993). If, however, high mountains stand above snow-bearing weather systems, the accumulation may decrease with altitude. The accumulation gradient can also be influenced by topography and by snow avalanching from adjacent valley sides.

The amount of mass gained or lost by a glacier in response to a change in the ELA depends on the *hypsometry* of the glacier. If, for example, a glacier has a large part of its area close to the ELA, a rising or lowering of the ELA will cause significant variations in mass. If, on the other hand, a minor proportion of the glacier is close to the ELA, ELA variations will have little effect.

Due to the wet-adiabatic lapse rate (altitudinal temperature change) of ca. 0.65°C per 100 m, ablation generally varies linearly with altitude. Ablation gradients are generally steepest where the summer temperature frequently rises above 0°C near the terminus, while temperatures higher up the glacier are below 0°C. Non-linear ablation gradients may be caused by altitudinal variations in cloudiness and humidity, proximity to rock faces, the amount of shading, aspect, and perhaps the most important factor, presence of debris on the glacier surface.

Accumulation and ablation gradients usually have different values, because they are controlled by different climatic variables. The ablation gradient is normally steeper than the accumulation gradient, showing an inflection at the equilibrium line altitude. The ratio between the two gradients is termed the balance ratio, given as

$$BR = b_{nb}/b_{nc} \qquad (4.3)$$

where b_{nb} and b_{nc} are the mass balance gradients in the ablation and accumulation zones, respectively. The balance ratio ignores any non-linearity which may exist in the respective mass balance gradients, but it is a useful

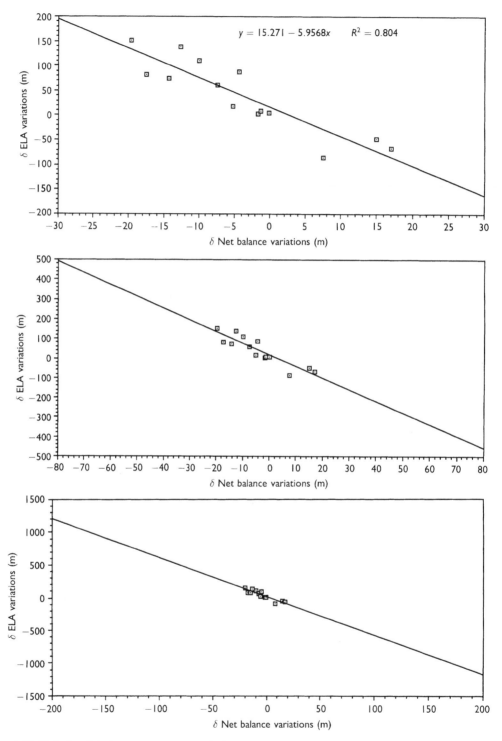

FIGURE 4.13 Relationship between net mass balance variations and ELA variations. The mass balance data are obtained from 14 glaciers in different climatic regimes worldwide (see Table 4.4)

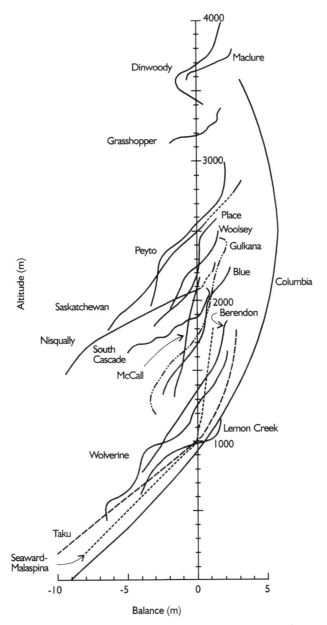

FIGURE 4.14. Annual mass balance gradients for glaciers in western North America. (Adapted from Benn and Evans, 1998)

parameter that summarizes the balance curve of a glacier. For 22 Alaskan glaciers, Furbish and Andrews (1984) found mean balance ratios of 1.8. A value of around 2 may be representative for mid-latitude maritime glaciers, while balance ratios for tropical glaciers may exceed 20 (Benn and Evans, 1998).

Several studies have tried to examine the complicated relationship between changes in glacier mass balance and climate variables. Chen and Funk (1990) correlated variations in mass balance of the Rhone Gletscher in Switzerland with climate records for the period 1882–1987. They found that most of the mass loss of the glacier was related to temperature increases, especially after 1940. Chen and Funk (1990) suggested that summer temperatures are in general more important than precipitation on mountain glaciers located in maritime climates. Nesje *et al.* (1995), however, demonstrated that both winter precipitation and summer temperature correlate with a 3–4 year lag in glacier front fluctuations of Briksdalsbreen, a western outlet of the semi-maritime Jostedalsbreen ice cap in western Norway. In the New Zealand Alps, Salinger *et al.* (1983) found that the retreat of the Stocking Glacier correlated with

monthly temperatures (two-year lag). On the opposite side of the water divide, however, variations of the Franz Josef Glacier were correlated by Hessell (1983) and Brazier *et al.* (1992) with precipitation changes. A similar effect was reported by Letréguilly (1988) and Pelto (1989) for glaciers in the coastal ranges of western North America. Negative cumulative mass balance of the South Cascade Glacier in the state of Washington, USA, is associated with reduced winter snowfall related to shifts in atmospheric circulation over the North Pacific Ocean and northern North America. Years of reduced mass balance on Peruvian and Bolivian

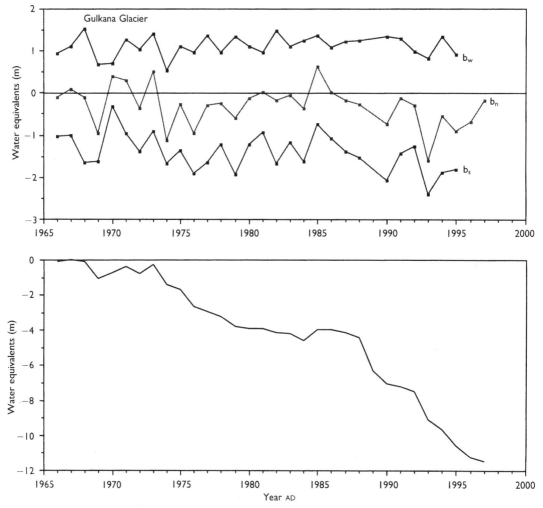

FIGURE 4.15 Annual winter (b_w), summer (b_s) (1966–1995), net (b_n) (1966–1997) (upper panel), and cumulative net balance (lower panel) at Gulkana Glacier. (Data from Jania and Hagen, 1996, and WGMS)

glaciers have been related to occurrences of the El Niño situation (Thompson *et al.*, 1984; Francou *et al.*, 1995). Most of the recent negative cumulative net balance of Lewis Glacier on Mount Kenya is related to air humidity effects on the energy balance on the glacier surface.

The formation of superimposed ice at the surface of high-Arctic glaciers is an important control on glacier mass balance (Woodward *et al.*, 1997). Increased temperatures are likely to reduce the extent and thickness of the superimposed ice, having a negative effect on the mass balance.

4.8.1 Long-term regional mass balance variations

Annual accumulation and ablation rarely balance, causing net mass gains or losses over a mass balance year. Variations in the net mass balance may average out over several years. In this case there is no long-term variation in the net mass balance. However, if the net mass balance is either positive or negative over several years, this will result in significant thickening or thinning of the glacier, respectively. Long-term trends in glacier mass balance are demonstrated by the cumulative net balance, or the running total of annual net balance.

A compilation of published mass-balance records from all over the world indicates that small glaciers appear to have been at equilibrium or shrinking slightly during the period between 1961 and 1990 (Cogley and Adams, 1998). For details about the different glaciers from which the mass balance records are obtained, see the World Glacier Monitoring Service's web page (www.geo.unizh.ch/wgms/index.html).

4.8.1.1 USA

Gulkana Glacier, Alaska. Gulkana Glacier (63°15'N, 145°28'W) is a valley glacier in the south-facing eastern Alaska Range. The glacier covers an area of 19.3 km^2, while the total drainage basin covers an area of 31.6 km^2. In 1965, mass balance studies by the United States Geological Survey (USGS) began. Annual mass balance variations are shown in Figure 4.15 (upper panel). The cumulative net

balance (Fig. 4.15, lower panel) shows that the glacier has decreased in thickness by 11.46 m water equivalents between 1965 and 1997. After 1988 the mass loss has accelerated. Regression analysis shows that the summer balance is the controlling factor for the net mass balance ($R^2 = 0.75$), while the winter balance is of negligible importance ($R^2 = 0.19$).

Wolverine Glacier, Alaska. Wolverine Glacier (60°24'N, 148°54'W) is a valley glacier in the Kenai Mountains, south central Alaska. The Kenai Mountains contain hundreds of smaller glaciers. The glacier and perennial snowfields cover about 72 per cent of the glacier catchment. From the 4 km wide accumulation basin, the glacier descends in a steep ice fall to an approximately 5 km long and 1.5 km wide valley glacier. The glacier has a maritime climate, despite being in the precipitation shadow of the Sargent Icefield, and the mass balance is considered to be quite representative of valley glaciers in the maritime part of Alaska. The annual mass balance variations are shown in Fig. 4.16 (upper panel). The cumulative net balance (Fig. 4.16, lower panel) shows that the glacier decreased in volume until 1979, when the glacier started to experience a positive net balance trend which lasted until 1988, after which the glacier has experienced significant mass loss. Between 1966 and 1997 the glacier volume was reduced by a layer corresponding to 7.68 m water equivalents. An evaluation of the mass balance factors influencing the net balance shows that the winter balance dominates ($R^2 = 0.69$), while the correlation between summer balance and net balance during the observation period is 0.21.

South Cascade Glacier. The annual net mass balance variations of the South Cascade Glacier (48°22'N, 121°03'W) between 1953 and 1997 are shown in Fig 4.17, p. 77 (upper panel). The cumulative net balance (Fig. 4.17, lower panel) shows a decreasing trend, especially after 1976. During the observation period the glacier has lost a surface layer corresponding to 22.88 m water equivalents. The steady-state ELA (ELA when $b_n = 0$) is close to 1910 m. The mean ELA for the periods 1971–1980 and 1986–1997 (22 years) was 1965 m, 55 m higher than the steady-state ELA.

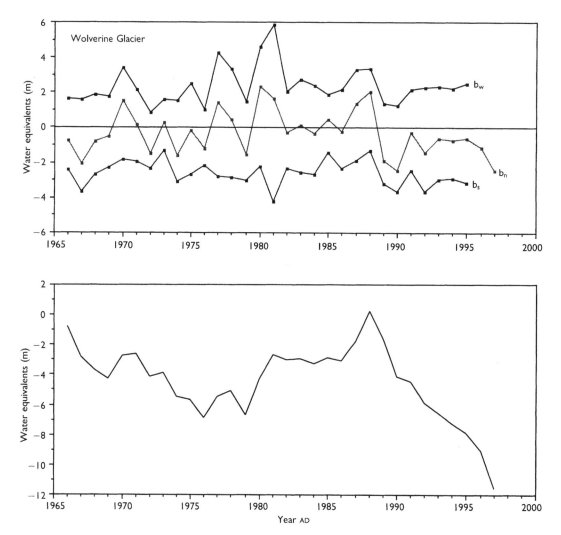

FIGURE 4.16 Annual winter (b_w), summer (b_s) (1966–1995), net (b_n) (1966–1997) (upper panel) and cumulative net balance (lower panel) at Wolverine Glacier. (Data from Jania and Hagen, 1996, and WGMS)

4.8.1.2 Canada

Devon Ice Cap. Mass balance measurements have been carried out on the northwest side of the Devon Ice Cap (75°20′N, 82°30′W) since 1961. The annual mass balance until 1995 is shown in Fig. 4.18, p. 78 (upper panel). For two years during the observation period (1976 and 1986) the summer balance was positive. The winter balance has been remarkably stable in contrast to the summer balance, which has shown great interannual variations.

The cumulative net balance (Fig. 4.18, lower panel) shows a decreasing trend; during the observation period the glacier has lost a surface layer corresponding to 1.795 m water equivalents. Regression analyses show that summer balance is the most important factor for the net balance ($R^2 = 0.88$), while there is no correlation between winter balance and net balance ($R^2 = 0.07$) for the observation period.

Place Glacier. Variations in annual net balance of Place Glacier (50°26′N, 122°36′W) between

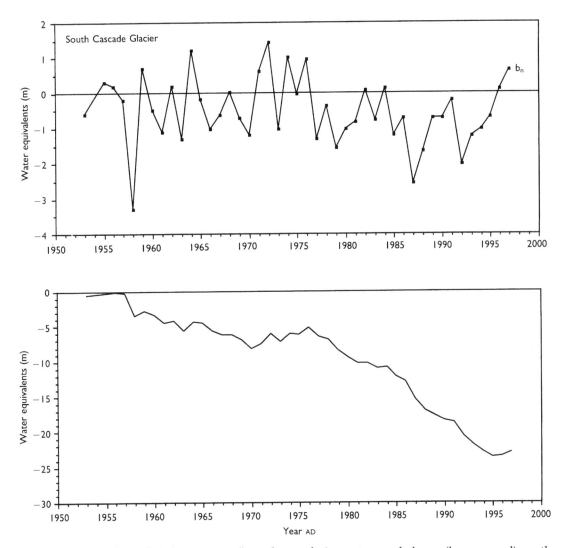

FIGURE 4.17 Annual net (b_n) (upper panel), and cumulative net mass balance (lower panel) on the South Cascade Glacier between 1953 and 1997. (Data from Jania and Hagen, 1996, and WGMS)

1965 and 1997 are shown in Fig 4.19, p.79 (upper panel). The cumulative net balance (Fig. 4.19, lower panel) shows a decreasing trend, which accelerated after 1976. The steady-state ELA (ELA when $b_n = 0$) is 2080 m, while the mean ELA during the observation period was 170 m higher at 2250 m.

Athabasca Glacier. Changes in areal extent, elevation and volume were calculated for Athabasca Glacier, Alberta, Canada, between 1919 and 1979 (Reynolds and Young, 1997).

During the study period, the glacier experienced a volume reduction of 2.344×10^8 m^3.

4.8.1.3 Norway

Ålfotbreen. The annual winter (b_w), summer (b_s), net (b_n) and cumulative net mass balance for Ålfotbreen (61°45'N, 5°39'E), a plateau glacier located close to the coast of western Norway, are shown in Fig. 4.20, p. 80, for the period 1963 to 1998. The steady-state ELA (ELA when $b_n = 0$) is ca. 1180 m, while the ELA

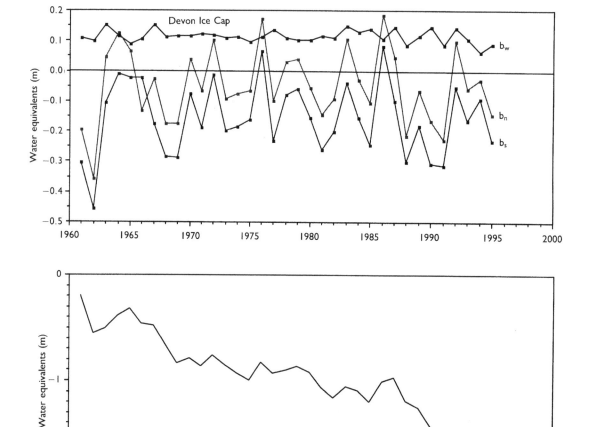

FIGURE 4.18 Annual winter (b_w), summer (b_s), net (b_n) (upper panel) and cumulative (lower panel) net balance at the Devon Ice Cap between 1961 and 1995. (Data from Jania and Hagen, 1996)

mean for the observation period is 35 m lower at 1145 m. The AAR when $b_n = 0$ is 0.52. Correlation analyses of the mass balance data show that winter precipitation is the main factor explaining the net balance variations ($R^2 = 0.76$), while the R^2 between b_n and b_s is 0.35.

Nigardsbreen. The annual winter (b_w), summer (b_s), net (b_n) and cumulative net mass balance of Nigardsbreen (61°43′N, 7°08′E), an eastern outlet glacier from Jostedalsbreen, are shown in Fig. 4.21, p. 81, for the period 1962 to 1998.

The steady-state ELA (ELA when $b_n = 0$) is 1560 m, while the mean ELA for the observation period is 65 m lower at 1495 m. Regression analyses of the mass balance data show that b_n and b_s are almost equally important for the net balance ($R^2 = 0.71$ and 0.70, respectively).

Hardangerjøkulen. The annual winter (b_w), summer (b_s), net (b_n) and cumulative net mass balance of Hardangerjøkulen (60°32′N, 7°22′E), a plateau glacier in central southern Norway, are shown in Fig. 4.22, p. 82, for the period 1963 to 1998. The steady-state ELA is

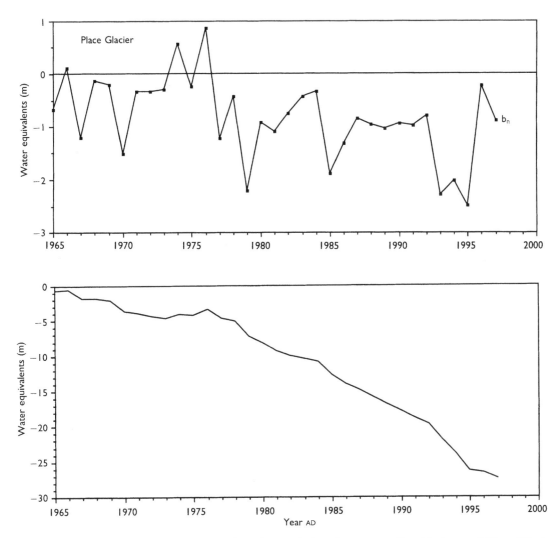

FIGURE 4.19 Annual net (b_n) (upper panel) and cumulative (lower panel) net balance at Place Glacier between 1965 and 1997. (Data from Jania and Hagen, 1996, and WGMS)

1680 m, while the mean ELA for the observation period is 75 m lower at 1605 m. Regression analysis of the mass balance data shows that the winter balance is the most significant factor explaining the net balance ($R^2 = 0.71$), while the determination coefficient between net balance and summer balance is 0.48.

Storbreen. The annual winter (b_w), summer (b_s), net (b_n) and cumulative net mass balance of Storbreen (61°34′N, 8°08′E), an east-facing glacier in western Jotunheimen (Liestøl, 1967), are shown in Fig. 4.23, p. 83, for the period 1949 to

1998. The steady-state ELA is 1720 m, while the mean ELA for the observation period is 30 m higher at 1750 m. Regression analyses of the mass balance data show that the summer balance is the dominant factor for the net balance ($R^2 = 0.66$), while the correlation coefficient between net balance and winter balance is 0.51.

Hellstugubreen. The annual winter (b_w), summer (b_s), net (b_n) and cumulative net mass balance between 1962 and 1998 on Hellstugubreen (61°34′N, 8°26′E) in eastern

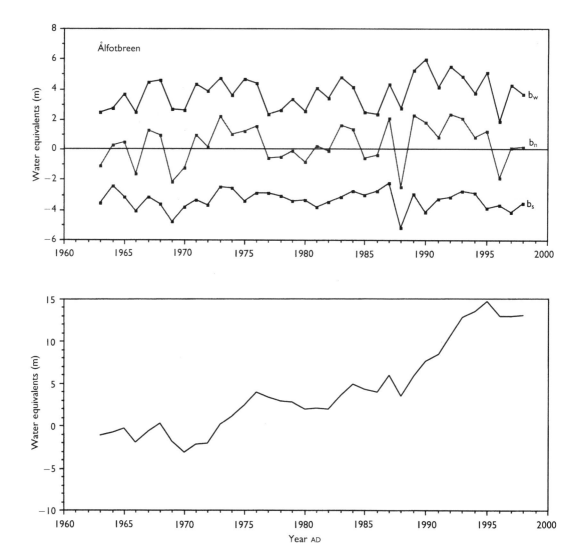

FIGURE 4.20 The annual winter (b_w), summer (b_s), net (b_n) (upper panel) and cumulative net mass balance (lower panel) on Ålfotbreen between 1963 and 1998. (Data from Kjøllmoen, 1998, and WGMS)

Jotunheimen are shown in Fig 4.24, p. 84. The mean ELA between 1963 and 1998 was 1900 m, which is 60 m higher than the steady-state ELA of 1840 m. Regression analyses of the mass balance data show that summer balance is the dominating factor ($R^2 = 0.80$), while the determination coefficient between the net balance and the winter balance is 0.34.

Gråsubreen. The annual winter (b_w), summer (b_s), net (b_n) and cumulative net mass balance

between 1962 and 1998 for Gråsubreen (61°39′N, 8°36′E), located in eastern Jotunheimen, central southern Norway, are shown in Fig. 4.25, p. 85. The steady-state ELA is 2060 m, while the mean ELA for the period 1962–1998 (except the year 1992) is 60 m higher at 2120 m. The AAR on the glacier at steady-state is 0.46. Regression analyses show that the summer balance is the main factor for the net balance variations ($R^2 = 0.80$), while the correlation between the net balance and the winter balance is 0.28.

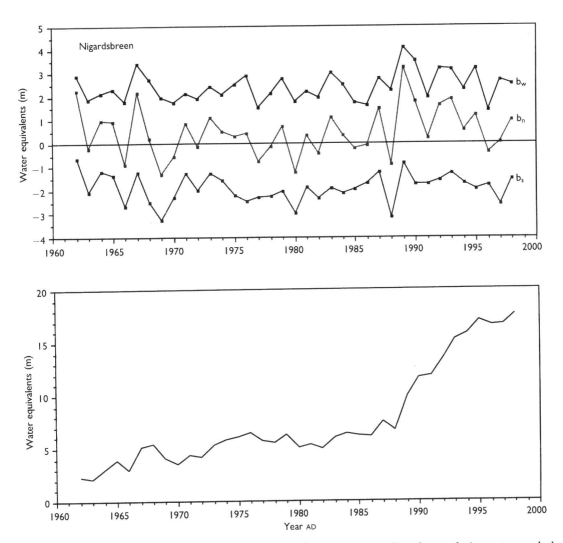

FIGURE 4.21 The annual winter (b_w), summer (b_s), net (b_n) (upper panel) and cumulative net mass balance (lower panel) on Nigardsbreen between 1962 and 1998. (Data from Kjøllmoen, 1998)

Engabreen. The annual winter (b_w), summer (b_s), net (b_n) and cumulative net mass balance between 1970 and 1998 for Engabreen (66°39'N, 13°51'E), a southwestern outlet glacier of Svartisen, are shown in Fig 4.26, p. 86. The mean ELA during the observation period was 1060 m, 100 m lower than the steady-state ELA of 1160 m. Regression analyses between winter balance and summer balance versus net balance shows correlation coefficients of 0.55 and 0.49, respectively.

4.8.1.4 Svalbard

Austre Brøggerbreen. The annual winter (b_w), summer (b_s), net (b_n) and cumulative net mass balance between 1967 and 1997 for Austre Brøggerbreen (78°53'N, 11°50'E) are shown in Fig. 4.27, p. 87. Ablation values show greater interannual variations than winter balance values. Because the summer ablation has been larger than the winter accumulation in all but two years in the observation

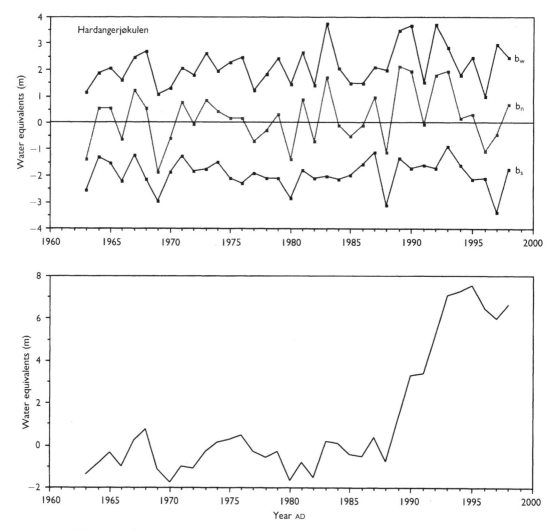

FIGURE 4.22 The annual winter (b_w), summer (b_s), net (b_n) (upper panel) and cumulative net mass balance (lower panel) for Hardangerjøkulen between 1963 and 1998. (Data from Kjøllmoen, 1998, and WGMS)

period, the glacier has experienced a steady volume decrease, with a total loss of 13.56 m during the observation period (Fig. 4.27, lower panel). The steady-state ELA on Austre Brøggerbreen is ca. 260 m, while the mean ELA during the observation period, except 1993, was 140 m higher at 400 m. The AAR on Austre Brøggerbreen at steady state is about 0.52. Regression analyses for b_w and b_s versus b_n show correlation coefficients of 0.08 and 0.76, respectively. Between the meteorological station at Ny-Ålesund and the glaciers only

5–6 km away, the correlation coefficient between the measured winter precipitation (September-June) at Ny-Ålesund and the measured snow accumulation from sounding profiles during the 14-year period from 1974/ 75 to 1987/88 was 0.63. The relatively poor correlation is mainly due to strong winds and snow-drifting (Hagen and Liestøl, 1990).

Midtre Lovénbreen. The annual winter (b_w), summer (b_s), net (b_n) and cumulative net mass balance between 1968 and 1997 for

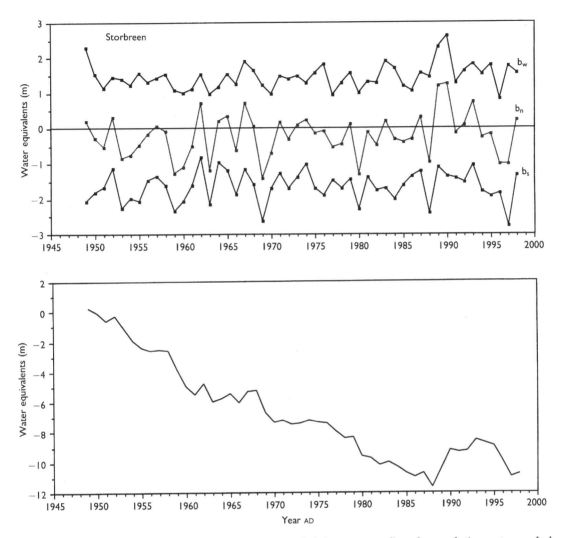

FIGURE 4.23 The annual winter (b_w), summer (b_s), net (b_n) (upper panel) and cumulative net mass balance (lower panel) on Storbreen between 1949 and 1998. (Data from Kjøllmoen, 1998, and WGMS)

Midtre Lovénbreen (78°53′N, 12°04′E) are shown in Fig. 4.28, p. 88. The cumulative net balance shows a steady decrease, with a total loss of 10.39 m during the observation period. The steady-state ELA on Midtre Lovénbreen is ca. 290 m, while the mean ELA for the observation period is 110 m higher at 400 m. The AAR at steady state is ca. 0.6. Regression analyses between winter balance/summer balance and net balance show correlation coefficients of 0.15 and 0.68, respectively.

4.8.1.5 Sweden

Storglaciären. The annual mass balance (upper panel) and cumulative net balance (lower panel) variations between 1946 and 1997 for Storglaciären (67°54′N, 18°34′E), northern Sweden, are shown in Fig. 4.29, p. 89. The cumulative net balance shows that the glacier reduced in volume until 1974. From 1988 the glacier has increased in volume. The steady-state ELA on Storglaciären is 13 m higher at

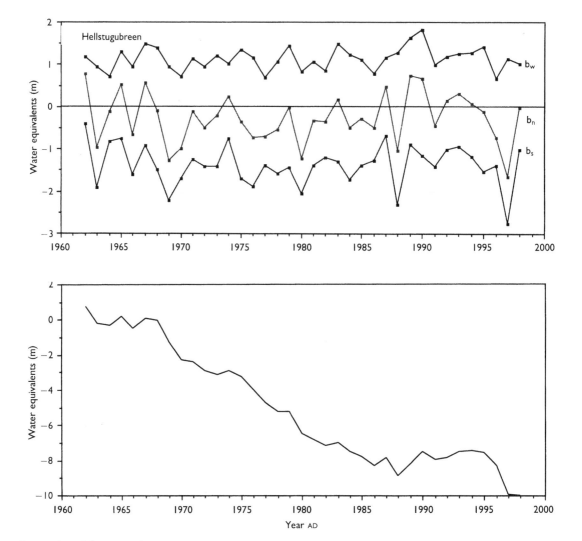

FIGURE 4.24 The annual winter (b_w), summer (b_s), net (b_n) (upper panel) and cumulative net mass balance (lower panel) for Hellstugubreen between 1962 and 1998. (Data from Kjøllmoen, 1998, and WGMS)

1460 m, while the mean ELA for the observation period (except the years from 1953 to 1959) was 1473 m. Regression analyses between winter balance/summer balance and net balance show correlation coefficients of 0.51 and 0.68, respectively. This demonstrates that the winter balance contributes significantly to the net mass balance variations on Storglaciären, as also pointed out by Raper *et al.* (1996).

Cumulative net mass balance variations for nine Norwegian (including Svalbard) glaciers and Storglaciären in northern Sweden (Fig. 4.30, p. 90) show that the maritime glaciers have increased their mass significantly, especially after 1988. The continental glaciers in southern Norway (Storbreen, Gråsubreen and Hellstugubreen), together with the Spitzbergen glaciers Austre Brøggerbreen and Midtre Lovénbreen, have decreased in mass.

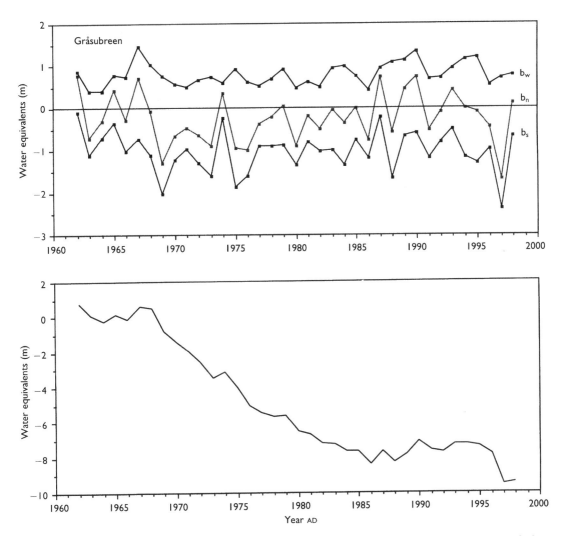

FIGURE 4.25 The annual winter (b_w), summer (b_s), net (b_n) (upper panel) and cumulative net mass balance (lower panel) for Gråsubreen between 1962 and 1998. (Data from Kjøllmoen, 1998, and WGMS)

Analysis shows that the net balance on the maritime glaciers is more influenced by the winter balance than by the summer balance, while the opposite is the case for the continental glaciers (Fig. 4.31, p. 90).

One source of interannual variability in the atmospheric circulation of NW Europe is the North Atlantic Oscillation (NAO). This oscillation is associated with changes in the westerlies in the North Atlantic and NW Europe (Hurrell, 1995; Hurrell and van Loon, 1997). After 1980,

and especially around 1990, the NAO tended to remain in an extreme phase and explained a substantial part of the observed temperature rise and increased precipitation during wintertime in NW Europe. Hurrell (1995) presented a NAO index for the period 1864–1995 based on air pressure gradients between Iceland and the Azores. The correlation between the NAO index and winter precipitation (Dec.–Mar.) in Bergen is 0.77. This is also reflected in the winter balance on glaciers in southern Norway

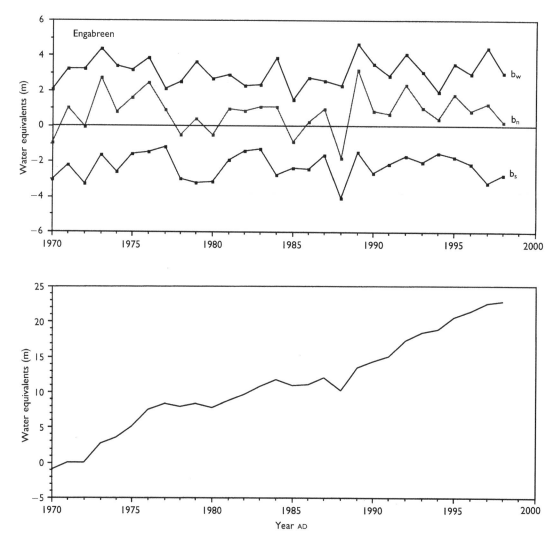

FIGURE 4.26 The annual winter (b_w), summer (b_s), net (b_n) (upper panel) and cumulative net mass balance (lower panel) for Engabreen between 1970 and 1998. (Data from Kjøllmoen, 1998, and WGMS)

(Fig. 4.32, p. 91); years with a high NAO index correspond to years of large winter balance, and *vice versa*.

4.8.1.6 Russia

Obruchev Glacier. The annual winter (b_w), summer (b_s), net (b_n) and cumulative net mass balance between 1958 and 1981 for Obruchev Glacier in the Urals are shown in Fig. 4.33, p. 92. From 1958 to 1981, the net mass balance loss was 3.22 m. The steady-state ELA is

520 m, while the mean ELA between 1960 and 1981 was 530 m. The AAR when the net balance on Obruchev Glacier is zero (steady state) is 0.50.

Maliy Aktru. The annual and cumulative net mass balance variations between 1962 and 1997 for Maliy Aktru (50°05′N, 87°45′E) are shown in Fig. 4.34, p. 93. The cumulative net balance shows significant variations; during the observation period the glacier lost a mass of 1.42 m water equivalents. The steady-state

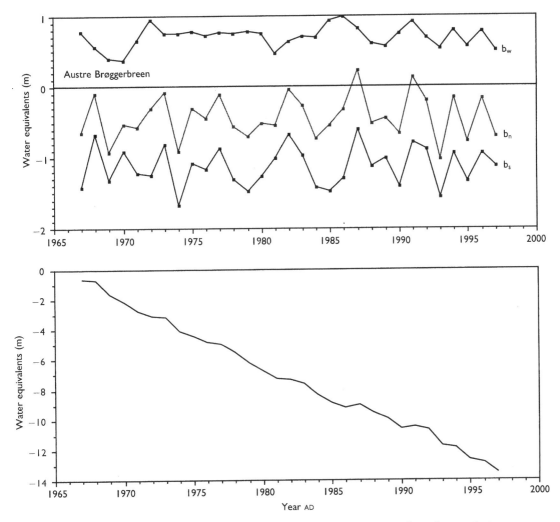

FIGURE 4.27 The annual winter (b_w), summer (b_s), net (b_n) (upper panel) and cumulative net mass balance (lower panel) for Austre Brøggerbreen between 1967 and 1997. (Data from Jania and Hagen, 1996, and WGMS)

ELA is 3140 m, while the mean ELA for the observation period was 3150 m. Regression analysis shows that the AAR at steady-state is 0.70.

4.8.1.7 Kirghizstan

Kara-Batkak. The annual and cumulative net balance between 1957 and 1997 for Kara-Batkak (42°08'N, 78°16'E) are shown in Fig. 4.35, p. 94. During the observation period the glacier reduced its mass by 17.95 m water

equivalents, most of which has occurred since 1972.

4.8.1.8 Kazakhstan

Ts. Tuyuksuyskiy. The annual and cumulative net mass balance variations between 1957 and 1997 for Ts. Tuyuksuyskiy (43°03'N, 77°05'E) are shown in Fig. 4.36, p. 95. The total mass loss during the observation period was 16.27 m water equivalents, most of which occurred after 1972. The steady-state ELA is

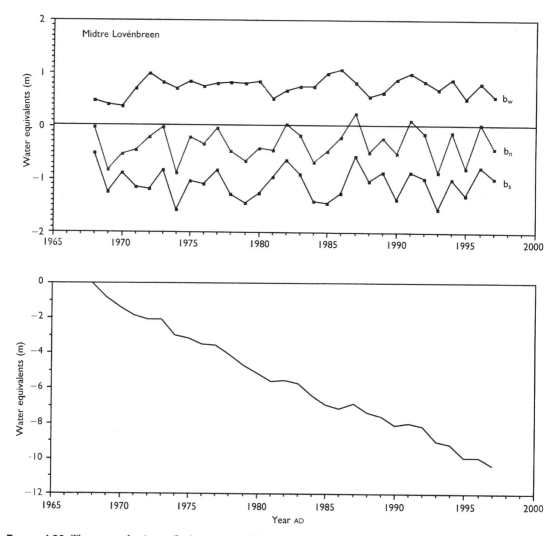

FIGURE 4.28 The annual winter (b_w), summer (b_s), net (b_n) (upper panel) and cumulative net mass balance (lower panel) on Midtre Lovénbreen between 1968 and 1997. (Data from Jania and Hagen, 1996, and WGMS)

3740 m, while the mean ELA during the observation period was 80 m higher at 3820 m. Regression analysis shows that the AAR under steady-state conditions is 0.54.

4.8.1.9 China

Urumqihe S. No. 1. The annual and cumulative net mass balance variations between 1959 and 1997 for Urumqihe S. No. 1 (43°05′N, 86°49′E) are shown in Fig. 4.37, p. 96. The cumulative net balance curve shows an accelerated net bal-

ance loss since the late 1970s. The total mass loss during the observation period was 6.0 m water equivalent. The steady-state ELA is 4030 m, while the mean ELA during the investigation period was 4050 m. Regression analysis shows that the AAR on the glacier under steady-state conditions is 0.50.

4.8.1.10 Austria

Hintereisferner. The annual and cumulative net mass balance variations on Hintereisferner

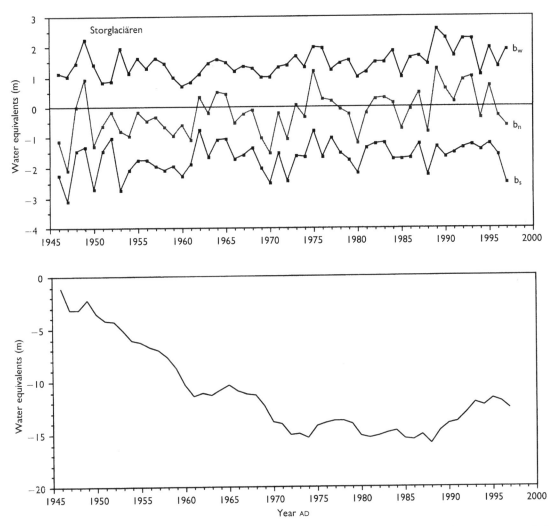

FIGURE 4.29 The annual winter (b_w), summer (b_s), net (b_n) (upper panel) and cumulative net mass balance (lower panel) on Storglaciären between 1946 and 1997. (Data from Holmlund *et al.*, 1996, and WGMS)

(46°48′N, 10°46E) in the Tyrol from 1953 to 1997 are shown in Fig. 4.38, p. 97. The cumulative net balance curve shows a decreasing trend from the beginning of the observation period until about 1980, when the mass loss increased significantly (total mass loss of 18.88 m water equivalents during the observation period). The calculated steady-state ELA on Hintereisferner is 2920 m, while the mean ELA during the observation period was 85 m higher at 3005 m. Regression analysis shows that the steady-state AAR is 0.66.

Equilibrium line altitudes have been reconstructed for Hintereisferner using temperature and precipitation records from 1859 to the present (Kerschner, 1997), by calibrating simple statistical models using observations of the ELA between 1964 and 1992. Correlation coefficients between observed and predicted ELAs are 0.91 for the glacial–meteorological model, and 0.98 for a multiple regression model, allowing backward extrapolation of the ELA from longer climatic records. The ELAs after the 1850 Little Ice Age maximum were rather

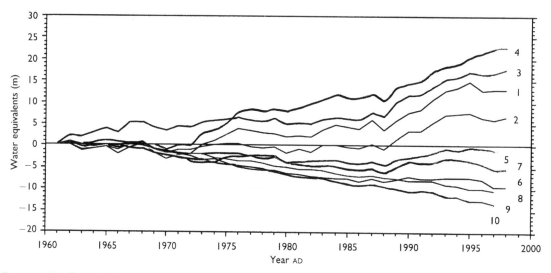

FIGURE 4.30 Cumulative net mass balance variations on Norwegian (including Svalbard) glaciers and Storglaciären in northern Sweden. 1, Ålfotbreen; 2, Hardangerjøkulen; 3, Nigardsbreen; 4, Engabreen; 5, Storglaciären; 6, Storbreen; 7, Hellstugubreen; 8, Gråsubreen; 9, Midtre Lovénbreen; 10, Austre Brøggerbreen.

high until the 1870s. The lowest ELAs, approximately 200 m lower than the 1850 mean, were recorded from 1912 to 1914. During the 1907–1926 period, the ELA at Hintereisferner was lower than the 1850 average for 13 out of 20 years, resulting in the extensive glacier advance during the 1920s in the Alps. From the early 1950s until 1980 the ELAs were generally

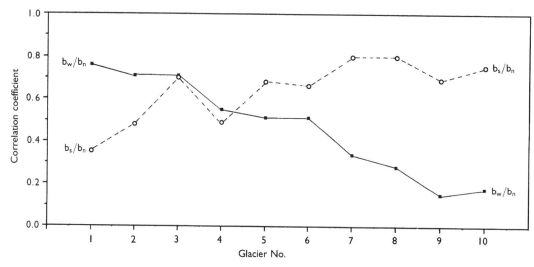

FIGURE 4.31 Correlation between winter balance (b_w) and net balance (b_n), and between summer balance (b_s) and net balance, for nine Norwegian (including Svalbard) glaciers and Storglaciären in northern Sweden, demonstrating that the net balance for the maritime glaciers is more influenced by the winter balance than by the summer balance, while the opposite is the case for the more continental glaciers. For numbering of the glaciers, see Fig. 4.30.

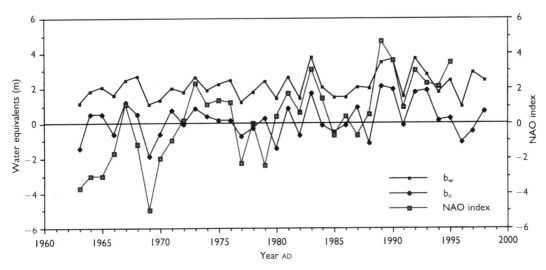

Figure 4.32 Winter balance (b$_w$) and net balance (b$_n$) on Hardangerjøkulen, central southern Norway (data from Kjøllmoen, 1998, and WGMS) and the North Atlantic Oscillation (NAO) index (Hurrell, 1995), demonstrating the close relationship between the NAO index and mass balance on a south Norwegian glacier. Similar numbers are used for both the NAO index (normalized sea-level pressure difference between the Azores and Iceland) and the mass balance data (in water equivalents).

lower. The period after 1980 has, however, been characterized by very high ELAs.

4.8.1.11 Switzerland

Silvretta. The annual and cumulative net mass balance variations for Silvretta (46°51′N, 10°05′E) from 1960 to 1997 are shown in Fig. 4.39, p. 98; the cumulative net balance curve shows significant interannual variations. The total mass loss since 1960 has been 0.3 m. The steady-state ELA on Silvretta is 2760 m, and the mean ELA during the observation period was also 2760 m. Regression analysis shows that the AAR under steady-state conditions is 0.57.

Griesgletscher. The mass balance measurements on Griesgletscher (46°26′N, 8°20′E) began in 1961 in connection with hydroelectric power construction (Funk *et al.*, 1997). The glacier covers an area of 6.2 km², and is about 5 km long. The glacier ranges in altitude from 2385 to 3375 m a.s.l. The annual net and cumulative mass balance variations between 1962 and 1997 are shown in Fig. 4.40, p. 99. The cumulative net mass balance curve shows significant

mass loss after 1979, the total mass loss being 9.81 m between 1961 and 1997. The steady-state ELA on Griesgletscher is 2830 m, while the mean ELA during the observation period (except for 1995) was 84 m higher at 2914 m.

4.8.1.12 France

Sarennes. The annual and cumulative net balance variations for Sarennes (45°07′N, 6°10′E) from 1949 to 1997 are shown in Fig. 4.41, p. 100. The cumulative net mass balance shows a decreasing trend, which, except for 1995, accelerated after 1984. The total mass loss during the observation period was 30.56 m water equivalents. Measurements of volumetric changes show that the Sarennes glacier was four times more voluminous 90 years ago, and five times bigger 150 years ago (Valla and Piedallu, 1997).

4.8.1.13 Kenya

Lewis Glacier. The annual and cumulative net balance variations on Lewis Glacier (00°09′S, 37°18′E) from 1979 to 1996 are shown in Fig. 4.42, p. 101. During the observation period,

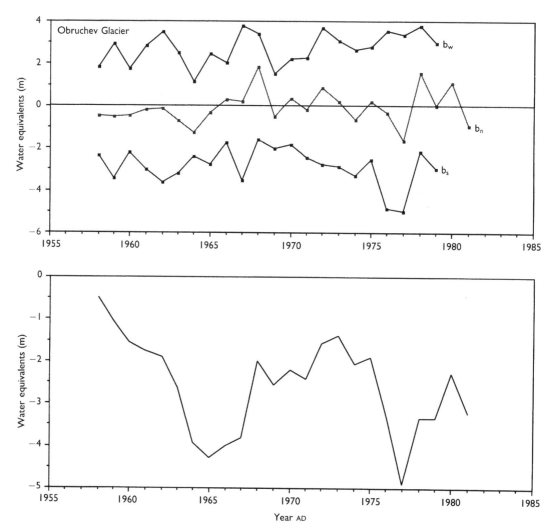

FIGURE 4.33 The annual winter (b_w), summer (b_s), net (b_n) (upper panel) and cumulative net mass balance (lower panel) on Obruchev Glacier in the Urals between 1958 and 1981. (Data from Jania and Hagen, 1996)

the total mass loss was 15.57 m water equivalent. Regression analysis shows that the steady state ELA is ca. 4810 m, while the mean ELA during the observation period was 75 m higher at 4885 m. The steady state AAR on Lewis Glacier is 0.38.

4.9 Frontal variations

The retreat and advance of glacier fronts has probably been used as a measure of climatic variations as long as humans have lived close to glaciated environments. Climate is constantly changing, with annual fluctuations superimposed on long-term trends. Such climatic changes are reflected in variations in the glacier extent. One example is the growth and decay of the Cenozoic continental ice sheets. Another example is the glacier advances during the 'Little Ice Age' and the subsequent frontal retreat. On a large scale, advances and retreats may be broadly synchronous. On a more detailed scale, however, the

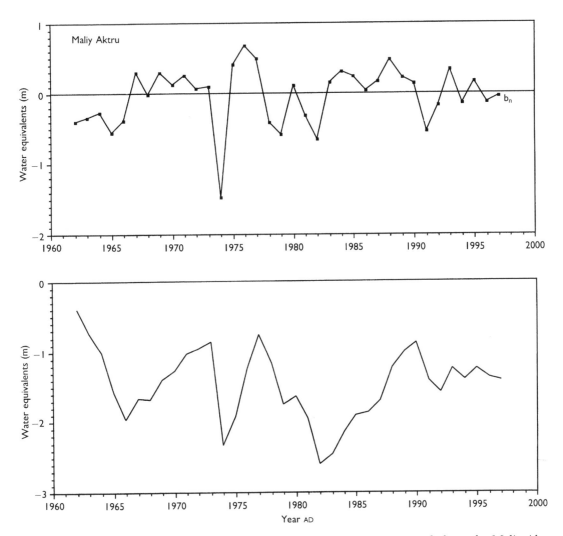

FIGURE 4.34 The annual (upper panel) and cumulative (lower panel) net mass balance for Maliy Aktru between 1962 and 1997. (Data from WGMS)

picture is more complex. In the same region, some glaciers may be advancing while others are retreating. Differences in local climates, aspect, size, steepness, and speed of individual glaciers may explain the different behaviour. In addition, the effect of a given climatic fluctuation on the glacier mass balance depends on the area–altitude distribution of the glacier. As a result, glaciers in the same area are likely to react differently, or at different rates, to the same mass balance variation (Paterson, 1994).

4.10 Response time/time lag

Advance and retreat of the glacier front normally lag behind the climate forcing because the signal must be transferred from the accumulation area to the snout. This is referred to as the *time lag*, or preferably the *response time*, which is longest for long, low-gradient and slowly moving glaciers, and shortest on short, steep and fast-flowing glaciers (e.g. Johannesson *et al.*, 1989; Paterson, 1994). Kinematic wave theory has been applied

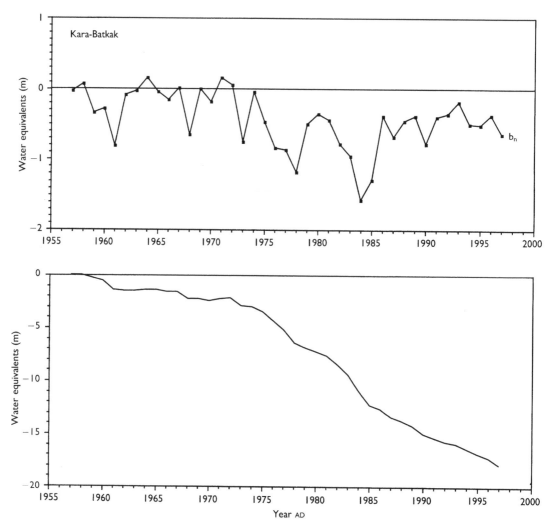

FIGURE 4.35 The annual (upper panel) and cumulative (lower panel) net balance for Kara-Batkak between 1957 and 1997. (Data from WGMS)

to calculating response times (Nye, 1960; Paterson, 1994). However, physically based flow models may help to determine the response times more precisely (van de Wal and Oerlemans, 1995).

The advance and retreat of glaciers are commonly the result of glacier mass balance changes. Theoretically, if the mass balance was constant for several years, the glacier would reach a steady state when the glacier size would remain the same, termed the datum state (Paterson, 1994). An increase in mass balance maintained for several years would lead to a new steady state. The altitudinal difference between the two glacier surface profiles increases steadily from the upper part and reaches a maximum at the position of the datum terminal position. Consequently, the head of the glacier does not change significantly, while the frontal part does, because the change in ice flux produced by the change in mass balance accumulates down-glacier. The response time is defined as the time a glacier takes to adjust to a change in mass balance (Paterson, 1994). The response time is the time the mass-balance perturbation takes to

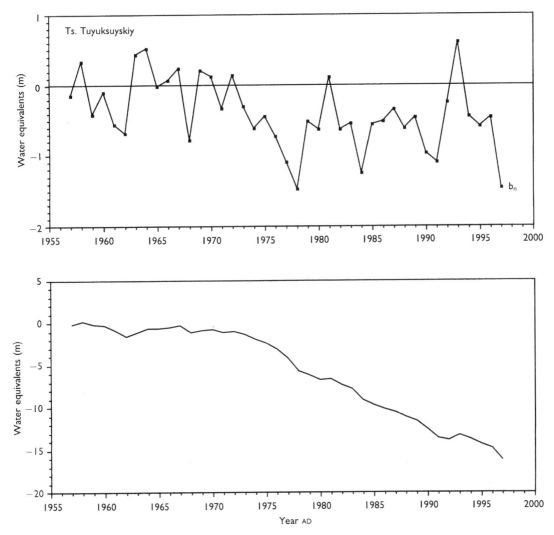

FIGURE 4.36 The annual (upper panel) and cumulative (lower panel) net mass balance variations for Ts. Tuyuksuyskiy between 1957 and 1997. (Data from WGMS)

remove the difference between the steady-state volumes of the glacier before and after the change in mass balance (Johannesson *et al.*, 1989). Glaciers in a temperate maritime climate with a thickness of 150–300 m, and an annual ablation at the terminus of 5–10 m, have estimated response times of 15–60 years. On the other hand, ice caps in Arctic Cascade, with a thickness of 500–1000 m and an annual ablation of 1–2 m, have estimated response times of 250–1000 years. The response time of the Greenland ice sheet is estimated to be around

3000 years (Paterson, 1994). It is, however, difficult to test these estimates, because glaciers are constantly adjusting to a complex series of mass-balance changes. Changes in mass balance are propagated down the glacier as kinematic waves, or more accurately a point, moving with a velocity different from the ice velocity (normally 3–4 times faster).

The relationship of glacier response to mass-balance changes is of great importance when climate variations are to be understood. McClung and Armstrong (1993) analysed two

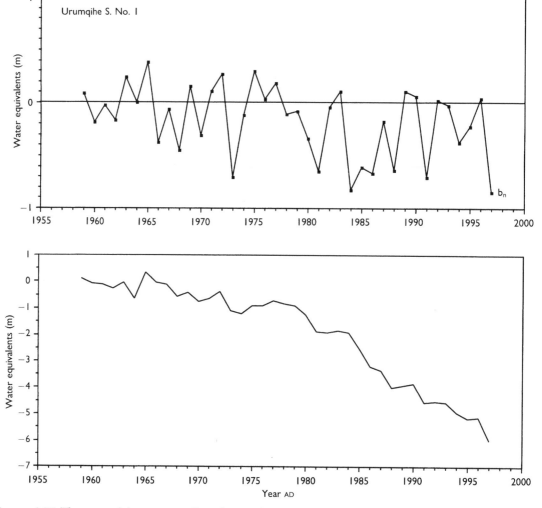

FIGURE 4.37 The annual (upper panel) and cumulative (lower panel) net mass balance variations for Urumqihe S. No. 1 between 1959 and 1997. (Data from WGMS)

aspects of this problem: (1) advance/retreat of the glacier front due to mass balance variations, and (2) cross-correlation of mass-balance data from two glaciers in the same climate zone. Their results indicate that the glacier terminus can respond quickly in accordance with expected minimum time-scales and that two adjacent glaciers may experience different annual mass balance and advance/retreat behaviour.

The complex dynamic processes linking glacier mass balance and length variations have only been studied numerically for a few

glaciers (e.g. Kruss, 1983; Oerlemans, 1988; Oerlemans and Fortuin, 1992; Greuell, 1992; Raper *et al.*, 1996). After a certain reaction time (t_r) following a change in mass balance, the length of a glacier (L_0) will start changing and finally reach a new equilibrium ($L_0 + \delta L$) after the response time (t_a). After full response, continuity requires that:

$$\delta L = L_0 * \delta b_t \qquad (4.4)$$

with $b_t = $ (annual) ablation at the glacier terminus. This means that, for a given change in mass balance, the length change is a function

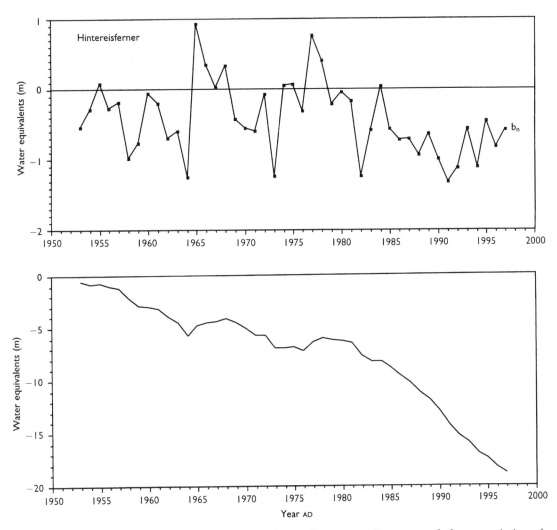

FIGURE 4.38 The annual (upper panel) and cumulative (lower panel) net mass balance variations for Hintereisferner in the Tyrol between 1953 and 1997. (Data from WGMS)

of the original length of a glacier, and that the change in mass balance of a glacier can be quantitatively inferred from the easily observed length change using estimates of b_t as a function of ELA and $\delta b/\delta H$, where H is altitude of the ice surface. The response time, t_a, of a glacier is related to the ratio between its maximum thickness (h_{max}) and its annual ablation at the terminus (Johannesson *et al.*, 1989)

$$t_a = h_{max}/b_t \qquad (4.5)$$

Corresponding values for valley glaciers are typically some decades. During the response time, the mass balance b will adjust to zero again so that the mean mass balance b is $0.5\delta b$ for a linear development. Cumulative length variation curves show that the smallest glaciers reflect annual changes in climate and mass balance with only a few years delay. Larger, more dynamic glaciers respond to decadal variations in climate and mass balance with a delay of several years, while the largest valley glaciers give smoothed signals of secular trends with a delayed response of several decades. For the latter two categories, the high-frequency, interannual front variations are filtered out.

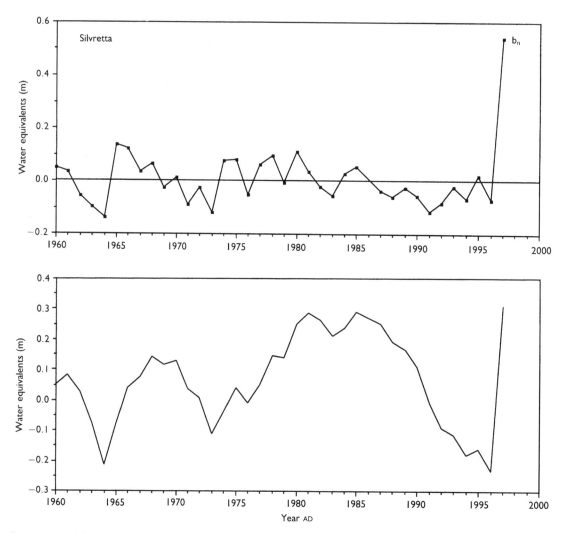

FIGURE 4.39 The annual (upper panel) and cumulative (lower panel) net mass balance variations for Silvretta from 1960 to 1997. (Data from WGMS)

4.11 Glacier movement

Information about glacier movement has been obtained from mathematical and numerical modelling, laboratory experiments, measurements in boreholes, observations in natural and artificial ice cavities, and interpretation of landforms and sediments in formerly glaciated regions.

Snow and ice are transferred from the accumulation area to the ablation area by glacier flow. Flow takes place as sliding, deformation of the ice, and deformation of the bed under

the glacier. The deformation and sliding due to gravity slowly transports snow and ice from the accumulation area and the interior of ice sheets to the marginal areas where ablation takes place. The rates of ice flow are related to the climatic input to the glacier and the geometry of the glacier catchment. For a glacier with no changes in size and shape, the ice flow through a cross section must be equal to the accumulation and ablation, illustrated by the wedge concept in Fig. 4.43, p. 102. Commonly, the maximum ablation occurs at the glacier snout, while the minimum

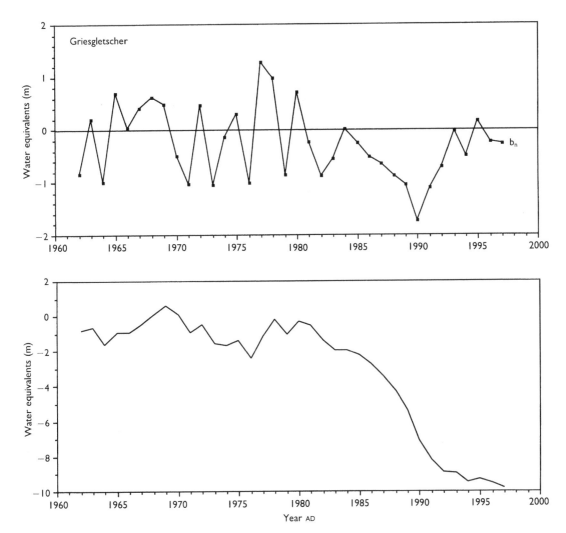

FIGURE 4.40 The annual net (upper panel) and cumulative (lower panel) mass balance variations on Griesgletscher between 1962 and 1997. (Data from Funk *et al.*, 1997, and WGMS)

ablation is recorded in the upper part of the glacier. The opposite is the case for accumulation. In order to be in a steady-state condition, glacier ice must move from the accumulation to the ablation area. Velocities calculated in this manner are termed balance velocities. As a consequence of this behaviour, ice velocities increase from the upper part, reaching a maximum at the equilibrium line. Below the equilibrium line the velocity decreases, reaching a minimum velocity at the glacier margin. Glacier velocities are highest on glaciers with

steep mass balance gradients, and where glacier ice from wide accumulation basins is focused into a narrow valley. The mass balance gradient at the equilibrium line is therefore an indication of the glacier activity. As a rule, glaciers in humid, maritime areas flow more rapidly than glaciers in arid, cold and continental regions. Consequently, glacier activity commonly increases with decreasing latitude and more maritime climates. However, glacier geometry highly influences glacier behaviour and therefore no simple relationship occurs

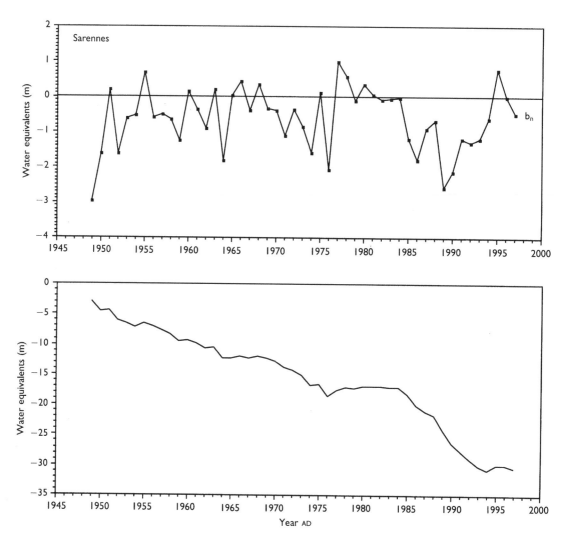

FIGURE 4.41 The annual (upper panel) and cumulative (lower panel) net balance variations on Sarennes from 1949 to 1997. (Data from WGMS)

between mass balance gradient and glacier discharge.

In the short term, significant deviations from the balance velocity occur as a result of the balance between driving and resisting forces being out of phase with variations in glacier mass balance. The driving forces are the stresses caused by the surface slope and the weight of the ice, and therefore influenced by variations in mass balance. The resisting forces are the result of the strength of the glacier ice, the glacier/bed contact and the

bed itself. The driving and resisting forces are kept in balance over longer periods by glacier flow. Glaciers therefore adjust their surface slope to produce driving forces sufficient to maintain the mass balance of the glacier. Temporary and significant changes in resisting forces may occur as a result of variations in water at the base of the glacier.

Stress and *strain* are important concepts for the understanding of glacier movement. Stress is a measure of how hard a material is pushed or pulled as the result of external

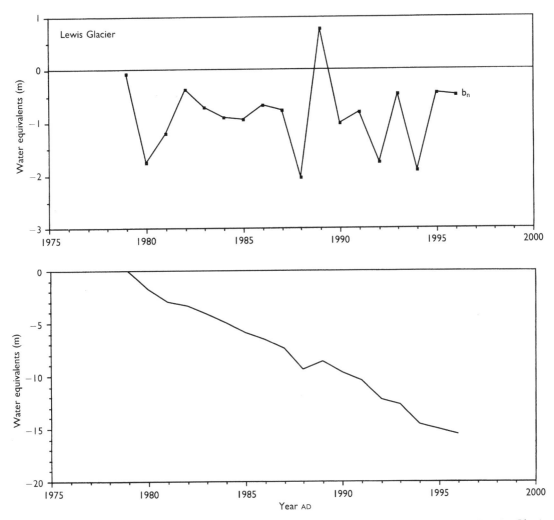

FIGURE 4.42 The annual (upper panel) and cumulative (lower panel) net balance on Lewis Glacier between 1979 and 1996. (Data from WGMS)

forces, while strain is a measure of the amount of deformation that takes place as a result of this stress. To understand stress, it is first useful to understand the concept of *force*, which is mass multiplied by acceleration. The downward force at the base of a glacier is the product of the mass of the overlying ice and the acceleration due to gravity, normally measured in newtons (N; $kg\,m\,s^{-2}$). Stress is defined as force per unit area. As a result, a given force acting on a small area will result in a greater stress than the same force acting over a larger area. The most commonly used units of stress are pascals ($1\,Pa = 1\,N\,m^{-2}$) and bars ($1\,bar = 100\,kPa$). The high stresses under the glacier make it more convenient to use kilopascals ($1\,kPa = 1000\,Pa$). The surface stress can be divided into two components: *normal stress* (the stress acting at right angles to the surface) and *shear stress* (the stress acting parallel to the surface). For normal stress, two opposing tractions either press the material together across the surface, termed *compressive stress*, or tend to pull the material apart across the surface,

(A)

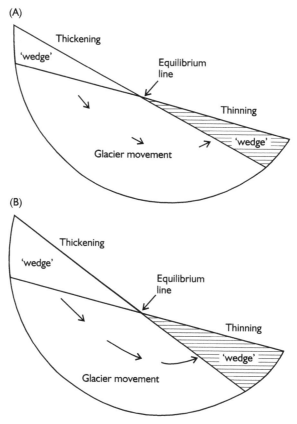

(B)

FIGURE 4.43 The wedge model of glacier flow. Glacier B has a steeper mass balance gradient than glacier A and therefore requires higher velocities to balance the mass gained and lost in the two wedges. (Adapted from Sugden and John, 1976, and Benn and Evans, 1998)

termed *tensile stress*. The tractions are parallel, but act in opposite directions for shear stresses. The shear stress *t* at the base of the glacier is a result of the weight of the overlying ice and the slope of the glacier surface, expressed as

$$t = rgh \sin a \qquad (4.6)$$

where *r* is the density of ice (ca. $0.9 \, \text{g cm}^{-3}$), *g* is gravitational acceleration ($9.81 \, \text{m s}^{-2}$), *h* is ice thickness in metres, and *a* is the surface slope. One implication of this formula is that the stress increases with ice thickness. Irregularities in the glacier bed may, however, result in stress variations that influence subglacial erosion and deposition. In addition, the pushing and pulling effect, termed *longitudinal stress*, is compressive where the velocity is slowing down and tensile where the velocity

is accelerating. Normal stress is greatest on the upstream side of subglacial bumps and lowest on the downflow side.

Strain, the change in shape and sometimes size of materials due to stress, can be divided into *elastic* (recoverable) strain and *permanent* (unrecoverable) strain. The degree of stress where permanent deformation occurs is termed the *yield* stress. Permanent deformation takes place as brittle failure (breaks along fractures) or ductile deformation (material is subject to flow or creep). Deformation involving volume changes is termed *dilation* (expansion or contraction), whereas deformation without volume changes is called *constant-volume* deformation. Strain types normally related to glacier flow are pure shear and simple shear (Fig. 4.44). Studies of glacier

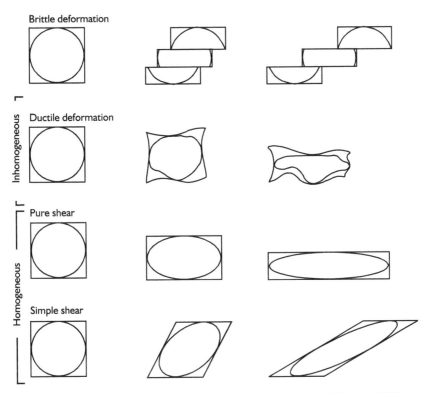

FIGURE 4.44 Styles of deformation in glaciers. (Modified from Benn and Evans, 1998)

movement commonly involve studies of *strain rate* (strain that occurs per unit time) and the *cumulative strain* (net amount of strain in a given period of time).

Strain my occur either by deformation of ice, deformation of the subglacial bed, or sliding at the interface between the ice and the bed. Surface movement on a glacier is the cumulative movement of one or all of these effects. Resistance to flow depends on several factors like temperature, debris content in the ice, bed roughness and water pressure.

Creep is ice deformation as a result of movement within or between ice crystals. Internal crystal movement can occur by sliding along cleavage planes, which are weakness lines related to crystal molecular structure, or movement along crystal defects. Movement between crystals leads to changes in shape or size by recrystallization at boundaries between crystals. Glen's flow law (Glen, 1955), first adapted

for glaciers by Nye (1957) is written:

$$e = At^n \qquad (4.7)$$

where e is strain rate, A and n are constants, and t is shear stress. Parameter A decreases with ice temperature; colder ice deforms less readily than ice at a higher temperature. The exponent n is normally close to 3 (Fig. 4.45). However, the orientation of ice crystals (ice fabric) and the presence of impurities (solutes, gas bubbles and solid particles) may cause strain rates different from those predicted from the flow law.

Fracture takes place when glacier ice can not move fast enough to allow the glacier to adjust its shape under stress. *Crevasses* are examples of fractures formed where ice is pulled apart by tensile stress. Tensile fractures are formed near the surface, because at depths greater than 15–20 m, the compressive force is larger than the tensile one. On temperate glaciers,

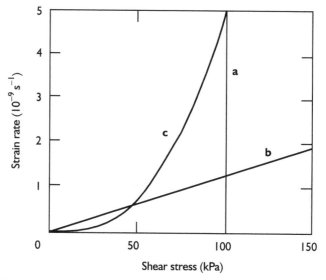

FIGURE 4.45 Relationships between stress and strain for different types of material: (a) perfectly plastic material; (b) Newtonian viscous material; (c) non-linearly viscous material (such as ice). (Adapted from Paterson, 1994)

crevasses are rarely deeper than 30 m, whereas on polar glacier, where the ice is stiffer than in temperate glaciers, the crevasses may be considerably deeper. Fractures can also form where ice is subject to compression where movement takes place along a shear plane.

Soft rocks and sediments underneath a glacier can experience strain as a result of stress by the glacier. This process of *subglacial sediment deformation* may explain certain types of unstable glacier behaviour and fast flow of some ice streams (e.g. Boulton and Jones, 1979). Studies of subglacial sediment deformation have been conducted in tunnels dug into glaciers, by placing instruments into boreholes, and by means of seismic sounding techniques. Studies have shown that subglacial material will not deform until a threshold stress is achieved, termed the *yield stress* or *critical shear stress*. The critical shear stress varies considerably, the most important factor being the effective normal pressure on the material from the overlying ice and sediment. The shear strength of a material is the sum of cohesion and intergranular frictional strength. Cohesion describes the forces keeping the material together, whereas frictional strength is the resistance of grains to sliding and grain

crushing. The deforming sediment layer is normally confined to the uppermost part of the bed, because the frictional strength increases downwards. Consequently, changes in the shear stress or in the porewater pressure will affect the thickness of the deforming sediment layer. The strain rates, increasing with increasing porewater pressure, commonly increase upwards, reaching a maximum value at the top of the deforming layer. Very high porewater pressure at the glacier–till interface, however, may cause decoupling and reduction of basal deformation rates.

Sliding is the movement between the glacier and the bed. The most important factors controlling the rate of basal sliding are adhesion due to freezing of ice to the bed, bed roughness, the quantity and distribution of water at the bed, and the amount of rock debris at the base of the ice. The rate of sliding is determined by the basal temperature, and effective sliding takes place when the ice is at the pressure melting point. The adhesive strength of frozen glacier beds is high even on a smooth surface. Until recently, therefore, sliding of cold-based glaciers was considered unlikely. Recent laboratory work and field evidence have, however, shown that slow basal sliding can take place

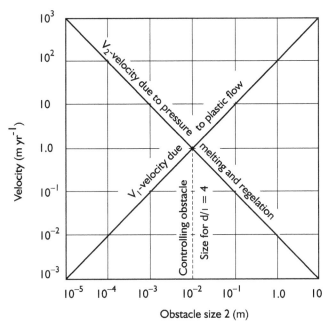

Figure 4.46 The relative effectiveness of regelation and creep for different obstacle sizes. (Adapted from Boulton, 1975)

underneath polar glaciers. Measurements below Urumqihe S. No. 1 in China showed sliding rates of $0.01 \, mm \, day^{-1}$ (annual sliding of less than 4 mm) at $-5°C$ (Echelmeyer and Wang, 1987; Echelmeyer and Zhong Xiang, 1987).

In nature, glacier beds are not smooth over wide areas, but have irregularities at different scales. The resistance to glacier movement around and above obstacles is termed *form drag*, which is an important factor controlling the rate of sliding. Two mechanisms of glacier movement over obstacles are *regelation sliding* and *enhanced creep*. During the regelation process, glacier ice slides over rough beds by melting on the upglacier side and refreezing on the downglacier side of the obstacle. Regelation sliding on the upstream side occurs from locally high pressures and lowering of the pressure melting point. The subglacial meltwater moves to the downglacier, low-pressure side of the obstacle, where it refreezes. The theory of regelation sliding suggests that this process is most effective when the latent heat on the downstream side of the obstacle is advected through the obstacle to melt the ice on the upglacier side (Weertman, 1964). The mechanism of

enhanced creep is the result of stress concentrations around the upglacier sides of obstacles, leading to accelerated sliding over the obstacle. The shape of the basal ice therefore changes continuously due to sliding. The creep rates are much lower in cold ice due to the temperature influence of the flow law (Glen, 1955). While regelation is most effective around small bumps, enhanced creep is most effective around large obstacles. A critical obstacle size (where neither of the two mechanisms is efficient and which represents maximum resistance to sliding) is found to be in the range 0.05–0.5 m (e.g. Kamb, 1970) (Fig. 4.46).

Liquid water at the glacier bed is fundamental for effective sliding, because of lack of adhesion at the bed and the need for water during the regelation process. Water pressure and the distribution of water are now recognized as the most important factors for modern sliding theory, explaining velocity fluctuations and glacier surge. The presence of a water film underneath glaciers moving over rock beds may accelerate sliding by submerging millimetre-scale surface irregularities. In addition, increased water pressure in cavities may have a similar effect. Field evidence suggests that water-filled cavities cause increased sliding velocity due to hydraulic jacking and uplift of the glacier (e.g. Willis, 1995). In other cases, glacier uplift may be related to variations in strain rate due to velocity variations and changes in longitudinal stress (e.g. Hooke *et al.*, 1989). The threshold water pressure for formation of cavities is termed *separation pressure*, which is high for short-wavelength, high-amplitude bumps and low for long-wavelength, low-amplitude surface irregularities (e.g. Willis, 1995).

Basal ice commonly contains rock debris, and where the debris is in contact with the bed, *frictional drag* will influence basal sliding. In general, the larger the drag is, the lower is the sliding velocity. Three models explaining subglacial friction have been suggested, called the 'Coulomb', 'Hallet' and 'sandpaper' models (Schweizer and Iken, 1992). The Coulomb friction model is based on the assumption that friction between rock particles in basal ice and a rigid bed is proportional to the normal pressure pressing the surfaces together (e.g. Boulton, 1979). In addition, the model infers that basal friction increases with ice thickness and is inversely proportional to basal water pressure. Hallet's (1981) friction model was based on the assumption that ice deforms completely around subglacial particles, and that contact forces are independent of ice thickness. The sandpaper model was introduced for debris-rich basal ice where particles are close or in contact. In this case, ice cannot flow around particles and the ice is the 'glue' holding the mass together. The ice rich in debris is deformable and therefore in contact with the bed. The drag force at the base of the debris-rich ice is a function of water pressure and the area of the bed occupied by cavities between particles in the basal ice layer. The sandpaper model is considered to be the best where the concentration of basal debris is more than about 50 per cent by volume, and the Hallet model the most appropriate when basal-debris concentrations are less than ca. 50 per cent by volume (Willis, 1995).

In a glacier, velocities vary commonly systematically with depth, the velocities being greatest at the surface at the centre of the glacier. The increase in velocity with height is most rapid near the bottom of the ice due to high strain rates. Near the surface, velocity increases more slowly with height due to lower strain rates in the upper ice layers, as illustrated by the velocity distribution of Athabasca Glacier in Canada (Harbor, 1992).

Positive mass balance in the accumulation zone of a glacier may be transmitted in waves of increased velocity, termed *kinematic waves* (e.g. Paterson, 1994), in which zones of increased velocity may travel downglacier at a rate approximately four times faster than the ice itself. Because the surface profile is constantly adjusting to mass-balance variations, kinematic waves are difficult to distinguish on non-surging glaciers (Benn and Evans, 1998).

Glaciers flow at a wide range of velocities. Cold-based glaciers with small balance velocities flow almost entirely by internal deformation, and velocities are relatively small (Echelmeyer and Wang, 1987). Wet-based glaciers, on the other hand, may reach high velocities. The fastest-moving glacier so far recorded is the Jacobshavn Isbræ, a tidewater outlet glacier on the western margin of the Greenland ice sheet, reaching maximum observed annual flow rates of 8360 m close to the calving margin (Echelmeyer and Harrison, 1990). For glaciers with annual velocities in the range of 10–100 m, basal shear stresses are relatively high (40–120 kPa), whereas the basal shear stresses under fast outlet glaciers and ice streams are considerably lower (10–30 kPa) (e.g. Paterson, 1994).

Velocity measurements carried out on Hintereisferner, central Alps, Austria (Span *et al.*, 1997), show three periods of accelerated flow between 1894 and 1994. These occurred around 1920, in 1940 and during the 1970s. The period of accelerated velocity around 1920 resulted in an advance of about 60 m. The mean annual advance increased from 30 m in 1914 to more than 120 m in 1919, and doubled during the accelerations in 1940 and 1980. The authors found that an increase in ice thickness alone cannot explain the enormous change in surface velocity, and that sliding may contribute to the total measured velocity during periods of accelerated ice flow.

4.12 Supraglacial ice morphology

Supraglacial structures reflect glacier formation, deformation and flow. Glacier ice exhibits
a wide variety of internal and superficial structures, such as *crevasses, icefalls, ogive banding,*
and *layering.* Crevasses form where ice is
pulled apart by tensile stresses that exceed
the strength of the ice, and they are commonly
oriented at right angles to the main stress
direction (e.g. Paterson, 1994). Crevasses are
therefore a reflection of the stress orientation
in a glacier. *Chevron* crevasses are linear features oriented obliquely upvalley from the glacier margins toward the centre of the glacier.
This type of crevasse forms in response to the
drag caused by the valley walls. Due to velocity differences between the centre and the
margin of the glacier, tensile stresses are
formed that pull the glacier from the margins
toward the central part, forming crevasses
oriented at 45° to the valley walls. *Splaying* or
marginal crevasses are formed due to compressive flow, causing the glacier to expand later-
ally. The crevasses are normally curved and
parallel to the flow direction toward the
centre. The crevasses normally bend outward
to meet the margins at angles of less than 45°.
Transverse crevasses form in a valley glacier
as a result of extending flow. Near the centre
of the glacier, the main tensile stress is parallel
to the glacier flow. Transverse crevasses therefore open up at right angles to the centre line.
Longitudinal crevasses are formed where the
lateral stress increases, for example, as a
result of widening of the valley glacier. On
the upper part of cirque and valley glaciers, a
randkluft is a fissure separating the glacier
from the rock wall. Such crevasses form as a
result of movement away from the rock back-
wall, and partly because of ablation adjacent
to warm rock surfaces. *Bergschrunds* are deep,
transverse crevasses near the heads of valley
and cirque glaciers.

When crevasses are formed, they will be carried by the glacier into areas of different stress
conditions. On valley glaciers and ice streams,
marginal crevasses tend to rotate because of
the higher velocity near the centre. Therefore,
rotated crevasses tend to close unless the

principal stress direction also rotates. When
crevasses close, they normally leave linear
scars, termed crevasse traces. These scars
may either contain layers of blue ice formed
by freezing of meltwater prior to closure, or
consist of thin layers of white, bubbly ice,
reflecting snow-filled crevasses. If the rate of
extension is small, the stretching is taken up
by the recrystallization of ice parallel to the
fracture, forming a tensional vein (Hambrey
and Müller, 1978).

Ice falls are steep parts of a glacier where the
flow is rapid and intermittent avalanches are
triggered by the collapse of ice blocks, termed
séracs, piling up cones of broken ice at the
base of the icefall. Where a glacier enters an icefall, the acceleration creates extreme extending
flow causing stretching and thinning under
tensile stresses. This leads to the formation of
a large number of crevasses, breaking the ice
up into séracs. At the base of icefalls, deceleration of the ice flow creates a zone of compressive flow, and crevasses are closed. In ice falls
there are commonly ice-free rock slabs or cliffs.

Ogives are regular bands or waves on the
surfaces of valley glaciers below icefalls. The
bands are convex downglacier due to a
higher velocity in the centre than at the margins. Banded ogives consist of alternating
light and dark ice, the dark bands consisting
of dirty ice and the light bands of more uniform bubbly ice. Wave or swell-and-wave
ogives consist of alternating ridges and
troughs. Ogives are considered to form
annually, one light–dark/ridge–trough pair
representing ice movement during one year.
Ogives may reflect seasonal variations in the
flow of ice down icefalls (Nye, 1958). Stretching
and thinning, in addition to the high concentration of crevasses, give the ice flowing
through the icefall a larger surface area than
the rest of the ice mass. During the summer
season, wind-blown dust and superficial
material will concentrate on the glacier surface
and form dark bands at the base of the icefall
when velocity decreases. Ice moving through
the ice fall in winter will, in contrast, collect
excess snow and form crests or light bands.

Glacier ice commonly exhibits different
types of layers or *foliation*. In the accumulation

area, layering may reflect annual cycles of snow accumulation. The layers are mostly parallel to the glacier surface and can be observed in the side walls of crevasses. The light bubbly ice represents the winter snow, while the thinner blue strata represent summer ablation surfaces. Dark layers in the ice are formed when wind-blown dust is concentrated during the summer melting. Foliation is formed in deep ice in the accumulation area and is observed in englacial and supraglacial ice in the ablation area. Foliation develops from layers or crevasse traces subject to high strain and/or rotation by glacier flow. Transverse foliation normally consists of crevasse traces downglacier from transverse crevasses and ice falls. Longitudinal foliation, on the other hand, is parallel to glacier flow and formed by the rotation of ice layers or crevasse layers, most commonly near the ice margin. Ice breccias form below intensively crevassed areas or ice cliffs where disintegrated ice is reconsolidated.

4.13 Glacier hydrology

Meltwater is an important component in glacier systems, and glacial meltwater exerts a strong influence on the hydrology of proglacial areas. Within and underneath glaciers and ice sheets, water affects glacier behaviour, controlling rates of glacier flow and influencing processes and rates of erosion and deposition. Surface and basal melting of glacier ice can produce large volumes of meltwater. In regions with low summer precipitation, but with glaciers in their catchment, glacial meltwater is an important source of water in the plant growing season, allowing cultivation of fields otherwise too arid for agriculture. Catastrophic floods from glaciers are, however, a recurrent threat in the Himalayas and in Iceland. In Scandinavia and in the European Alps, glacier meltwater has also been used for hydroelectric power production. The movement of glaciers and ice sheets is also strongly influenced by pressure and distribution of meltwater.

Water in a glacier drainage system may originate from melting of ice and snow, rainfall, runoff, or sudden release of stored water

(Fig. 4.47). The runoff from glacier catchments varies significantly both annually and seasonally. Surface melting increases with rising air temperature, solar radiation, and rainfall when air temperatures are relatively high. Englacial and subglacial melting can also take place as a result of frictional heat caused by deforming and/or sliding ice, geothermal heat, and pressure melting when the glacier moves over topographic irregularities.

Water drainage through a glacier is influenced by the permeability of the ice. In the sense of glacier hydrology, we talk about *primary permeability* (permeability of intact ice and snow) and *secondary permeability* (related to the size and distribution of tunnels and crevasses). Primary permeability is commonly high for snow and firn where the air spaces between the snow crystals are linked. In solid ice, on the other hand, the primary permeability is low, because the air bubbles are more or less isolated from each other. However, in ice at the pressure melting point, water can penetrate through systems of interconnected lenses and veins between the ice crystals (Benn and Evans, 1998, and references therein). Below the pressure melting point, intact ice is impermeable. Most of the water draining through a glacier is related to secondary permeability, the water flowing through a system of conduits of varying length and diameter. Englacial conduits in temperate glaciers are formed and maintained by melting from circulating air and flowing/standing water. Polar glaciers do not normally have englacial conduit systems, so most of the surface water drains supraglacially. Crevasses may, however, take meltwater down to the basal ice layers which can be at the pressure melting point.

The water flow is determined by variations in the *hydraulic potential*, which is a measure of the available energy at a particular time and place. For water flowing on the surface, the hydraulic potential depends on the elevation only, and the water flows downslope. The flow of *englacial* and *subglacial* meltwater is, however, more complicated because the hydraulic potential depends on altitudinal differences and water pressure. The hydraulic

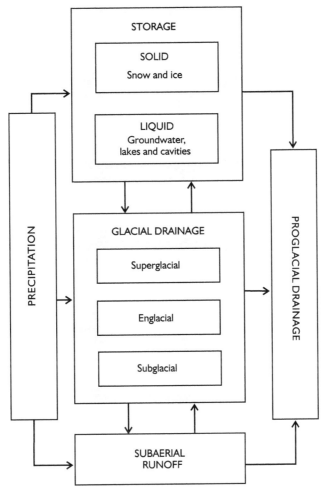

FIGURE 4.47 Water sources and pathways in glaciated catchments. (Adapted from Benn and Evans, 1998)

potential in an englacial or subglacial conduit is dependent on the shape and size of the conduit, elevation and water pressure. The water pressure may vary between *atmospheric pressure* (pressure of the open air) and *cryostatic pressure* (pressure of the weight of the overlying ice). The difference between the water pressure and the ice pressure is termed the *effective pressure*, an important property for sub- and englacial drainage.

The flow direction in an englacial and subglacial drainage network is mainly controlled by variations in the hydraulic potential; the water follows the hydraulic gradient, flowing from areas of high potential towards areas of low potential. The gradient depends on the surface slope, and to a minor extent, the slope of the water-filled conduit. The *equipotential surfaces* (planes connecting points with the same potential) therefore rise downglacier with a gradient of about ten times that of the ice surface. Water draining freely through a glacier will flow at right angles to the equipotential surfaces.

Hodson *et al.* (1997) compared estimates of suspended-sediment yield and discharge from two glacier basins in Svalbard. Austre Brøggerbreen (12 km^2) is almost entirely cold-based, whereas Finsterwalderbreen (44 km^2) is dominated by basal ice at the pressure melting point. Specific suspended-sediment yields from Finsterwalderbreen of 710–2900 t km^{-2} a^{-1} were

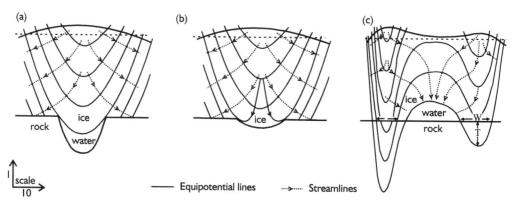

FIGURE 4.48 Subglacial water body formation. (a) A deep depression beneath an ice dome; (b) a shallow depression beneath an ice dome; (c) a subglacial cupola formed beneath a depression on the ice surface. (Adapted from Nye, 1976)

more than an order of magnitude larger than at Austre Brøggerbreen ($81–110\,t\,km^{-2}\,a^{-1}$). The difference is explained by the influence of the thermal regime of the meltwater drainage system and the sources of suspended sediments.

If water flow is prevented, water may be stored in subglacial, supraglacial, englacial or ice-dammed lakes. Such water bodies are commonly temporary, expanding and contracting in response to glacier fluctuations, glacier dynamics or volcanic activity. Subglacial water bodies vary significantly in size; the largest modern reservoirs known from radio echo-sounding under the Antarctic ice sheet are up to $8000\,km^2$ in area. In areas with low hydrological potential surrounded by areas where the hydrological potential is high, water may accumulate in subglacial ponds (Ridley *et al.*, 1993). Hydrological gradients will in this case cause the water to drain towards the water reservoir. Figure 4.48 shows situations where subglacial ponding can occur. In case (a), water ponds within a bed depression beneath an ice dispersal centre. When the bedrock depression is shallower than the shape of the equipotential contours (b), ponding will normally not occur. In the case of (c), the glacier bed is nearly horizontal and there is a central ice depression. In this case, the hydrological potential beneath the depression is lower than the surrounding areas. Water will commonly flow towards the bed beneath the depression and form a

dome-shaped upper surface along an equipotential surface.

An effective mechanism for the production of subglacial lakes is the melting caused by subglacial volcanic activity. The western part of Vatnajökull in Iceland is located above an active part of the mid-Atlantic spreading ridge. Glacier ice resting on the Grimsvötn caldera melts and form a supraglacial depression. The lake system at Grimsvötn empties about every six years, in most cases catastrophically. Such an event normally involves the release of up to $4.5\,km^3$ of water, with a maximum discharge of ca. $50,000\,m^3\,s^{-1}$. A large subglacial eruption in the Grimsvötn caldera in the autumn of 1996 caused an enormous *jökulhlaup*. The Icelandic term jökulhlaup is used for periodic or occasional release of large amounts of stored water in catastrophic floods. Jökulhlaups may be caused by sudden drainage of ice-dammed lakes, overflow of lake water, or the growth and collapse of subglacial reservoirs. The best-documented jökulhlaups are reported from Iceland. During these flooding events, the large sandur plains of Skeiderarsandur and Myrdalssandur are totally flooded.

Where glaciers or ice sheets form a barrier to meltwater drainage, water will be stored to form an ice-dammed lake. At valley glaciers, ice-dammed lakes may form in ice-free tributary valleys blocked by the glacier in the main valley. In other cases, tributary valley glaciers may block water drainage. A third case may be that ice-dammed lakes form at the junction between two valley glaciers. One of the largest ice-dammed lakes was Lake Agassiz (2 million km^2), which formed during the deglaciation of the Wisconsin event (e.g. Teller, 1995). In Siberia, northwards draining rivers were dammed by the Eurasian ice sheet(s), forming huge ice-dammed lakes.

4.14 Calving glaciers

Several mechanisms of *glacier calving* have been described; however, the fundamental control and relationship with calving rate is poorly known. This makes it difficult to explain the lack of climatic sensitivity of different glaciers, in particular the order-of-magnitude difference in calving rate between tidewater and lake-calving glaciers (Fig. 4.49). The mechanism for 'normal' slab calving includes the development of an overhang before failure occurs as a result of deformation of the glacier close to the ice cliff. Minor normal faults or englacial shear bands have been observed and explained as forward bending of the cliff. Profiles of calving cliffs normally show rotation of the cliff profile about the base.

Kirkbride and Warren (1997) used repeated photographs and field surveys to reveal the mechanism of ice-cliff evolution at Maud Glacier, a temperate glacier calving in a lake in New Zealand. Their study showed that calving is cyclic: (1) waterline melting and collapse of the roof of a sub-horizontal notch at the cliff foot; (2) calving of ice flakes from the cliff face leading to a growing overhang from the waterline upward and cracks opening from the glacier surface; (3) calving of slabs as a result of the developing overhang, returning the cliff to an initial profile; (4) seldom subaqueous calving from the ice foot.

The relationship between climate and calving glaciers is not straightforward, and it is rarely possible to draw reliable conclusions about climate variations from calving glaciers (Warren, 1992). Iceberg calving leads to instability in the glacier, causing the glacier to oscillate asynchronously with climatic variations

and with other calving and non-calving glaciers. The rate of calving is primarily controlled by water depth. Calving dynamics are, however, poorly understood. The dynamics seem to be different from temperate to cold/polar glaciers, and between grounded and floating fronts.

The physical processes that control calving rates are complex, poorly understood, and not yet quantified (Bahr, 1995; van der Veen, 1995). There is, however, a strong linear correlation between calving speed and water depth (Fig. 4.49) at grounded, temperate calving fronts. This relationship can be applied in both freshwater and marine environments, but the slope coefficient is some 15 times greater at tidewater termini (Warren *et al.*, 1997). The instability introduced to glaciers by calving introduces heterogeneous glacial responses to climate (e.g. Motyka and Begét, 1996) through the complicated interaction of the calving front with topography and effective water depth (Sturm *et al.*, 1991).

Glacier Uppsala, a freshwater calving glacier in southern Patagonia, has been retreating since 1978 (Naruse *et al.*, 1997). Subsequent to a significant retreat of approximately 700 m in 1994, the recession seems to have ceased in

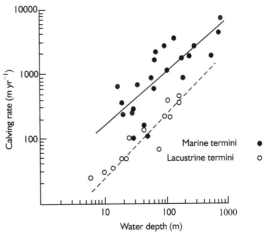

Figure 4.49 The relationship between water depth and calving rates at the margins of tidewater glaciers (solid dots) and freshwater glacier termini (open circles). (Adapted from Benn and Evans, 1998)

1995. A thinning rate of 11 m per year was recorded close to the front between 1993 and 1994. They found that temperature alone could not explain the reason for this thinning. It was therefore suggested that calving was the main reason for this extensive thinning at the glacier terminus.

Most Patagonian glaciers are retreating rapidly (Casassa et al., 1997). Glacier O'Higgins, a freshwater calving glacier, has experienced the largest retreat from 1945 to 1986. Climate warming, combined with the detachment of the glacier front from an island, are believed to be the main causes of the retreat (Casassa et al., 1997).

4.15 Surging and tidewater glaciers

A *surging* glacier commonly shows little sign of unusual activity for several years. Then it starts to move rapidly and the glacier surface, especially in the lower part, is transformed from a fairly smooth surface into deep crevasses and ice pinnacles. The frontal part may move several kilometres in a few months or years before it suddenly stops. Ice velocities during most surges are 10–100 times those in normal glaciers. Observed velocities vary from about 100 m per day over short periods, to 200–500 m yr^{-1} for glaciers surging for one to two years. Before a surge, the glacier upstream from the terminus normally consists of stagnant ice covered by debris. During a surge, this part of the glacier becomes dynamically active and thickens. The actively moving front may, however, not proceed beyond the area of the previous stagnant ice mass. In this case it is misleading to speak of a net glacial advance.

Surges seem to start when the glacier reaches a critical surface profile. At the end of a surge, on the other hand, the profile of the glacier surface is much lower than for a steady-state glacier. During a surge, large quantities of ice are moved from an upper to a lower area of the glacier. In the upper part of the glacier, the pre-surge ice surface may be visible along the valley sides as remaining ice with steep ice walls down to the surging ice surface. The high velocities observed in surging glaciers cannot be explained by ice deformation, but by basal sliding and/or bed deformation. Therefore, the basal ice must be at the pressure melting point during the surge. Between the surges, on the other hand, some surging glaciers seem to be frozen to the substratum. The rapid decrease in ice velocity during the later stages of surges have been accompanied by floods in the glacier meltwater streams. Large quantities of silt during these outbursts may indicate that effective subglacial erosion can take place during surges.

Surging glaciers have commonly been related to subpolar or polythermal glaciers of a variety of sizes and types. The majority of the surging glaciers have been described from western North America near the Alaska–Yukon border, to the St. Elias Mountains, the Alaska Range, and to parts of the Wrangell and Chugach Mountains. In Iceland, the main, gently sloping outlet glaciers from Vatnajökull are known to surge. In Asia, surging glaciers have been mapped in the Pamirs, Tien Shan, the Caucasus, Kamchatka, and the Karakoram. Surging glaciers have also been reported from the Chilean Andes, Greenland, Svalbard, and from Arctic Canada (see Paterson, 1994, and references therein). A compilation of data from surging glaciers (Paterson, 1994: 359) suggests that most surges occur at fairly regular intervals. There is no evidence of surging ice sheets, although a surging behaviour may take place in the West Antarctic ice streams.

Recent research aimed at understanding the behaviour of surge-type glaciers has generally taken two different approaches. The first approach has been to carry out extensive process studies at individual glaciers, focusing on the nature of the surge mechanism, surge cycles and basal processes (e.g. Raymond and Harrison, 1988). The second type of approach has been to study several surge-type glaciers in a wider region (e.g. Dowdeswell et al., 1991).

Tidewater glaciers (glaciers standing with their front in a fjord or in the sea) in the fjords of Alaska have retreated significantly during the last two centuries. The most spectacular changes have taken place in the Glacier Bay, one of the glaciers having retreated by about 100 km since the late eighteenth century. In contrast, some tidewater glaciers have been advancing for the last hundred years. The reported advance rates are, however, significantly less than retreat rates (20–40 m yr^{-1} and 200–1700 m yr^{-1}, respectively).

Bakaninbreen began surging in 1985/86, forming a surge front where fast-moving surge ice met non-surging ice (Porter *et al.*, 1997). Up to 1995 the surge front moved 6 km downglacier. Yield strengths calculated for the basal sediments at Bakaninbreen range between 16.6 and 87.5 kPa. The estimates of basal shear stress suggest that sediments upglacier of the surge front will be actively deforming, while only limited deformation will take place downglacier of the surge front.

During the 1985/86 surge of Bakaninbreen, a surge front up to 60 m high was formed (Murray *et al.*, 1997). Shear zones and thrust faults were formed in association with the forward movement of the surge front. A ground-penetrating radar showed several subglacial and englacial debris layers reflecting thrusting by the glacier.

Finsterwalderbreen (35 km^2), a polythermal glacier in southern Spitzbergen, last surged around the beginning of the twentieth century (Nuttall *et al.*, 1997). Surface elevations have been measured since 1898, showing thinning and frontal retreat. The accumulation area, however, is gradually building up. At present, velocities increase from about 1 m per year at the front to 13 m per year at the equilibrium line. At the bergschrund, the velocity is about 5 m per year. Radar profiles indicate that the glacier is at the pressure melting point at the base, which is also supported by hydrological studies showing high suspended-sediment loads. The authors suggest that the glacier may be building up towards another surge.

The Columbia Glacier is a large, temperate tidewater glacier that calves into the Columbia Bay in Alaska. Venteris *et al.* (1997) found regular seasonal cycles in speed and stretching rate. These cycles continued after the glacier retreated off the shoal at the end of the fjord around 1983. They suggested that seasonal change in subglacial water is the main controlling factor of the velocity of the glacier. The

change in terminus position (extensive retreat) seems to be linked to thinning caused by longitudinal extension, as proposed by van der Veen (1996).

Five surges of Variegated Glacier in Alaska indicate that they all terminated with their surge fronts in the terminal lobe of the glacier, and that different surges penetrated into the terminal lobe by different amounts (Lawson, 1997). Of five studied surges, the one that occurred in 1905–6 penetrated furthest into the terminal lobe. A surge in 1964–65 affected a greater proportion of the glacier than the 1947–48 and the 1982–83 surges.

4.16 Reconstruction of ice-surface profiles and calculation of basal shear stress

From studies of modern glaciers, the *basal shear stress* is known to be related to the surface profile and thickness of glaciers, expressed by the equation:

$$t_b = rgh \sin a \qquad (4.8)$$

where t_b is basal shear stress, r and g the ice density and gravitational constant, respectively, a the surface slope of the glacier, and h the glacier thickness (Paterson, 1994). The value of t_b (the basal shear stress) can be calculated from measurements of ice thickness h and surface slope a. Most calculated values of basal shear stress vary between 50 and 100 kPa, suggesting that glacier ice behaves like a perfectly plastic material with a yield stress of 100 kPa.

For evaluation of the effects of sides of valley glaciers and ice streams on stresses and velocities, the shape factor F was introduced, given by the half-width/ice thickness on the centre line. Thus, in a valley glacier the valley walls support part of the weight of the glacier, inducing a basal shear stress on the centre line less than its value for a wide channel. A factor in the range of 0.5–0.9 may therefore be inserted in the denominator on the right-hand side of the equation (4.8).

Measurements along flow-lines of modern ice sheets show that most of their profiles can

TABLE 4.5 Values of basal shear stress (t_b) and the corresponding k value

t_b (kPa)	k
5	1.1
10	1.5
20	2.1
30	2.6
40	3.0
50	3.4
60	3.7
70	4.0
80	4.3
90	4.5
100	4.7

be expressed by different *parabola* and *ellipse* equations. Glaciologically this is explained by the plastic properties of ice being the dominating factor determining the shape of glaciers and ice sheets (Paterson, 1994). For perfect plastic ice sheets the profile can be expressed by the equation:

$$h^2 = (2t_0/rg)(L - x) \qquad (4.9)$$

where t_0 is shear stress and $L - x$ the distance from the ice margin along a flow line.

In constructing ice-sheet profiles, a simple approach in the case of a perfectly plastic ice sheet can be used:

$$H = k\sqrt{L} \qquad (4.10)$$

where H is the altitude of the glacier surface at distance L from the margin, both values in metres. The constant k refers to basal shear stress and may be as high as 4.7 (basal shear stress of 100 kPa). Calculated basal shear stress in ice sheets, however, varies between 0 and 100 kPa, with an average around 50 kPa, giving k = 3.4. Values of basal shear stress (kPa) and the corresponding k values are given in Table 4.5.

The shear stress can be calculated when the glacier thickness and surface gradient are known and when the glacier profile is mapped by means of lateral moraines and/or trimlines. One property of the parabola equations is that the altitudinal difference between glacier profiles becomes smaller with increasing distance from respective glacier termini.

Small irregularities (less than 1 m high) in the glacier sole effectively reduce basal sliding. If the glacier is at the pressure melting point, a thin water film is commonly present between the glacier sole and the substratum. If the water film is thick enough to reduce the friction between the ice and the ground, the basal shear stress will be reduced significantly. This will in turn lead to low-gradient glacier profiles and ice thickness may be halved at the transition between cold/polar and temperate ice. An important requirement for such a reduction of the basal shear stress is that the water is not draining away.

In polar/cold glaciers, all or most of the glacier movement occurs as internal deformation. When the temperature in the ice mass is below the pressure melting point, the ice does not deform so easily as it does when it is on the pressure melting point. The terrain relief is important for glaciers with low-gradient surface profiles. The overall gradient of the landscape is, however, of minor importance for the surface profile. Deformation of underlying sediments reduces the basal shear stress and cause low-gradient ice-sheet and glacier profiles. An important factor for subglacial sediment deformation is high sediment pore-water pressure. Rapid retreat of fjord glaciers may be explained by break-up and calving when the glacier front starts to float. A rapid retreat of fjord glaciers normally causes steep and dynamically unstable glaciers. The precipitation distribution over an ice sheet can influence surface profiles. Of peculiar importance are variations in the relationship between accumulation and ablation (the nivometric coefficient) on different parts of the glacier.

Glacier profiles based on parabolic equations require that the glacier is in equilibrium. However, such a balance between ablation and accumulation will not normally occur on large ice masses. The ice thickness will remain more or less constant if the volume of ice flowing through the glacier is of similar magnitude to the net balance. If this relationship is not maintained, local accumulation and ablation effects can influence the shape of the glacier profile significantly.

As the continental shelf off Norway and large areas of mainland Scandinavia covered by the Late Weichselian ice sheet were underlain by deformable sediments, we must take into consideration the recent information on subglacial processes in reconstructing the thickness of the ice sheet. As a result, glacier flow in the present shelf area could have been a product of deformation of basal till and sediments as well as deformation of ice. The ice thickness may therefore have been influenced by the yield stress of the subglacial sediments.

Modelling of the Laurentide and Scandinavian ice sheets that incorporates the effects of possible substrate deformation indicates a central ice-dome surface about 400–800 m lower than those based on the assumption of a rigid bed, perfectly plasticity and a yield stress of 100 kPa (Boulton *et al.*, 1985; Fisher *et al.*, 1985; Clark *et al.*, 1996). Low-gradient ice-sheet profiles have been reconstructed in the southern Baltic region, over the present continental shelf area in the North Sea, along the fjord areas of southern Norway (Nesje and Sejrup, 1988), and in NW Scotland (e.g. Ballantyne *et al.*, 1998). These gently sloping

ice-sheet profiles can best be explained by subglacial sediments deforming at low yield strength, or perhaps by water-lubricated sliding, or a combination of the two. The evidence for thin, low-gradient ice margins in the area covered by the Scandinavian ice sheet is in accordance with the reconstruction of former lobes and outlet glaciers of the Late Wisconsin Laurentide ice sheet, which indicate similar thin ice margins (e.g. Klassen and Fisher, 1988).

Licciardi *et al.* (1998) presented a series of numerical reconstructions of the Laurentide ice sheet during the last deglaciation from 18,000 to 7000 yr BP, evaluating the sensitivity of the ice-sheet geometry to subglacial sediment deformation. Their reconstructions assumed that the Laurentide ice sheet flowed over extensive areas of water-saturated, deforming sediments. Their reconstructions suggest a relatively thin and multidomed ice sheet.

In addition, several factors like the land–sea distribution, precipitation pattern, equilibrium line altitude and ice-sheet frontal fluctuations, and isostatic movements, influenced the geometry of the Late Weichselian Scandinavian ice sheets (Nesje *et al.*, 1988; Nesje and Dahl, 1990).

5

Glacier variations

5.0 Chapter summary

Chapter 5 summarizes the evidence of pre-Quaternary glaciations and the glacial/interglacial cycles during the Quaternary, from both terrestrial and marine records. In addition, evidence of late Cenozoic glacier and climate variations is presented from the Arctic, Eurasia, Northern Europe, the Alps, North and South America, New Zealand and Antarctica. A section is devoted to Late Glacial glacier and climate variations in NW Europe. Records of Holocene glacier and climate variations are presented from Canada, the USA, the Arctic, Greenland, Iceland, New Zealand, South America, Antarctica, Mount Kenya and other East African mountains, the Alps, China, Scandinavia, the Tatra Mountains and Jan Mayen. Neoglacial and Little Ice Age glacier variations are presented from Iceland, Scandinavia, France, the Alps, Eurasia, China, North and South America, Greenland, East Africa and the Pyrenees, with a synthesis, comparison and global summary of local glacier variations. One section deals with glaciers in relation to environmental change and the human race. Finally, models of Late Quaternary climate and ice-sheet evolution are presented.

5.1 Pre-Quaternary glaciations

A general global temperature curve (Fig. 5.1) suggests that favourable conditions for glaciation occurred during at least five periods of Earth's history. Most probably, however, the Earth has experienced some kind of glaciation during the entire time span, but glaciation may have varied from continental ice sheets to minor ice caps or alpine glaciers. For pre-Quaternary glaciations, it is important to pay attention to continental drift and mountain building in order to explain the timing and geographical distribution of major glaciations. Since the development and acceptance of the theory of plate tectonics in the early 1960s, several attempts have been made to apply the theory to ancient glacial deposits, most of which involved correlating glaciation on a particular continent with periods when the continent was at high palaeolatitude (Table 5.1). This was successfully applied to glacial deposits of the southern continents (Gondwanaland) during the Palaeozoic (e.g. Crowell, 1978).

Evidence for pre-Quaternary glaciations is dominated by tillites, mixtites or diamictites as sediments and lithified rocks. The oldest evidence for glaciation is from the Precambrian. The Huronian glaciation dates to the Lower Proterozoic (approx. 2000–2500 million years ago) and is represented by ca. 12,000 m thick sediments in the Lake Huron region of Canada. Evidence for Late Precambrian glaciation is available on almost all of the present continents. Extensive glaciations also occurred during the Palaeozoic (570–230 million years ago). In Africa, located over the South Pole during Ordovician times, there is reliable evidence for Ordovician glaciation. The ice sheet complex there may have been some 20 times larger than the present Antarctic ice

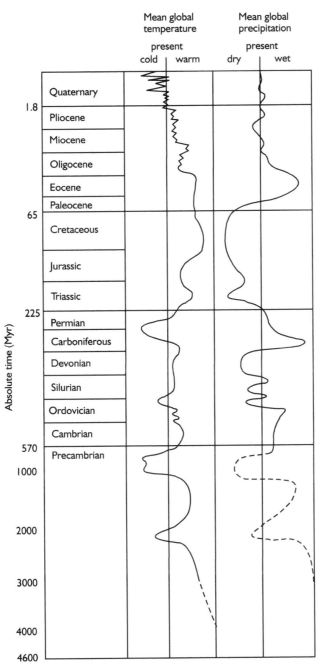

FIGURE 5.1 General global temperature and precipitation curves – note variable timescale. (Adapted from Bradley, 1985)

sheet. Ordovician glaciations have also been reported from southern Africa, South and North America and Scotland. In South America and southern Africa there is fragmentary evidence for Devonian glaciation. The Permo-Carboniferous glaciation of Gondwanaland, centred over what is now southern Africa and Antarctica, had perhaps the greatest

TABLE 5.1 Timing of some Cenozoic global events related to glaciation (Adapted from Herman *et al.*, 1989)

Time (Ma)	Tectonic events	Climatic events
0.9–0	Himalayan, Alpine orogeny	Major glacial–interglacial cycles
0.9	Orogeny peak	
1.6	Tibetan, Himalayan, and Sierra Nevada orogenies	
2.4–2.7		First major northern hemisphere ice sheets
3.5	Uplift of Panama Isthmus	Gradual temperature decline
	Opening of Bering Strait	Southern hemisphere lowland glaciation
5.5	Isolation of Mediterranean Sea	
6		Strong global cooling
		Expansion of Antarctic ice sheet
12–14		Major Antarctic build-up?
16–18	Subsidence of Iceland–Faeroe ridge	
14–18	Renewed Himalayan Tibetan orogeny	
24	Drake Passage opens	Increased glaciation of Antarctica
		Intensification of global climatic gradients
36	Greenland separates from Eurasia. Tasman seaway opens	Antarctic continental glaciation
35–37	Major Himalayan and Alpine orogenies	Cooling at high and low latitudes
		Mountain glaciation in Antarctica

impact of all the pre-Quaternary glaciations. The ice sheet reached the coasts of eastern South America, southern Africa, southern Australia and southern India. During its maximum extent, the ice sheet covered about twice the area covered by the present Antarctic ice sheet.

The early development and build-up of the glaciers took place during the late Tertiary. The initial glacier formation may have been caused by tectonic uplift of the Tibetan Plateau, the Himalayas, and the western Cordillera in North America (e.g. Ruddiman *et al.*, 1989). The development history of the Antarctic ice sheet is a matter for debate. Offshore sediments have been interpreted to indicate glaciation during the last 40 million years (Hambrey *et al.*, 1989, 1992), while others have suggested glacier formation 5–10 million years ago (e.g. Drewry, 1978). The Antarctic ice sheet has fluctuated since its initial formation, and some scientists have even suggested that for short periods it may have melted away (e.g. Barrett *et al.*, 1992). This view has been challenged, for example by Denton *et al.* (1993), presenting geomorphological, geochronological and palaeoclimatological evidence indicating that

the East Antarctic ice sheet has been fairly stable during the last 4.4 million years. In the Gulf of Alaska, there is a record of late Tertiary glaciation represented by glaciomarine sedimentation from the late Miocene into the Quaternary (Eyles *et al.*, 1991).

According to a recent study by Maslin *et al.* (1998), the onset of northern hemisphere glaciation began in the late Miocene with a significant build-up of ice in southern Greenland. Progressive intensification of glaciation, however, did not seem to have begun until 3.5–3 million years ago, when the Greenland ice sheet expanded to include northern Greenland. Those authors suggested that the Eurasian Arctic and Northeast Asia were glaciated at about 2.74 million years ago, Alaska at 2.7 million years ago, and North East America at 2.54 million years ago. Tectonic changes (uplift of the Himalayan and Tibetan Plateau, deepening of the Bering Strait, and the emergence of the Panama Isthmus) were suggested to have been too gradual to be responsible for the speed of northern hemisphere glaciation. The authors therefore suggested that tectonic changes brought global climate to a critical threshold and that relatively rapid variations in the Earth's orbital parameters (insolation) triggered the northern hemisphere glaciation.

5.2 Glacial/interglacial cycles during the Quaternary

The Quaternary is conventionally subdivided into *glacials* and *interglacials*, with further subdivision into *stadials* (shorter cold periods within interstadial or interglacial stages) and *interstadials* (shorter mild episodes within a glacial phase). Glacial stages are normally considered as cold phases of major expansion of glaciers and ice sheets. Interglacials are considered as warm periods when temperatures were at least as high as during the present Holocene interglacial. These terms, however, lack precision and are difficult to apply. Still, they are widely used among Quaternary scientists. A subdivision into a coherent scheme of glacial and interglacial stages for the stratigraphic record from the land areas of the northern hemisphere has been extremely difficult. Therefore a system based on the deep ocean oxygen isotope record has come into use. The marine isotope signal is mainly controlled by the global volume of terrestrial ice, and fluctuations in the isotopic signal can be looked upon as a record of glacial and interglacial fluctuations. A system of *isotope stages* has therefore been developed. Counting from the most recent, each isotope stage has been given a number, even numbers reflecting stadial glacial/cold stages, while odd numbers denote interstadial interglacial/warm phases. The deep sea oxygen record is global and therefore geographically consistent. During the last 800,000 years there have been about ten interglacial and ten glacial stages, while during the past 2.5 million years the total number of isotope stages exceeds 100, corresponding to between 30 and 50 glacial/interglacial cycles (Ruddiman *et al.*, 1989; Ruddiman and Kutzbach, 1990) (Fig. 5.2).

Palaeoclimatic records from the deep sea floor show that during the last 3 million years more than a hundred global mild and cool oscillations took place with a larger amplitude than those recorded during the Holocene.

Variations of dry and wet climates in the Chinese Loess Plateau and in the loess belt of central Europe have a similar frequency (Kukla, 1987). The periodicity of past major climate variations agrees well with the orbital perturbations. The amplitude of the orbital cycles and of the marine oxygen isotope variations is relatively uniform. However, several of the cold climate episodes recorded on land seem more severe than others (Chinese loess units L1, L2, L5, L6 and L9; central European loess cycles B, C, F, H and J; corresponding to oxygen isotope stages 2, 6, 12, 16 and 22 and the Weichselian, Saalian and Elsterian ice sheet advances).

5.2.1 The terrestrial record

The most detailed system of glacial and interglacial episodes has been developed for Northern Europe, the British Isles, North America and the Alps. Most terrestrial evidence has, however, been destroyed by subsequent glaciations, and the terrestrial geomorphological and stratigraphic record of former glaciations is therefore incomplete. The last ice age in northern Europe, the *Weichselian*, is considered to be equivalent to the *Devensian* in Britain, the *Würmian* in the European Alps, and the *Wisconsin* in North America. The last interglacial, the *Eemian*, *Ipswichian*, *Riss-Würmian* and *Sangamon* may all correlate with one another. In Europe, a sequence of stadials and interstadials has been recognized from areas (periodically) beyond the limit of the Weichselian glaciations. The terrestrial record of glacials and interglacials prior to the last interglacial has been more difficult to resolve. An inter-regional attempt at correlation is shown in Box 5.1. Recent work in the European Alps, northern Europe, North America and Britain suggests that several of the established stages contain several separate warm and cold episodes.

In central Europe and China, loess deposits and intercalated soil horizons correlate closely with glacial fluctuations. Long lacustrine sequences with associated sediments together with animal and plant remains have the potential to be correlated directly with deep sea records (e.g. Kukla, 1977, 1989).

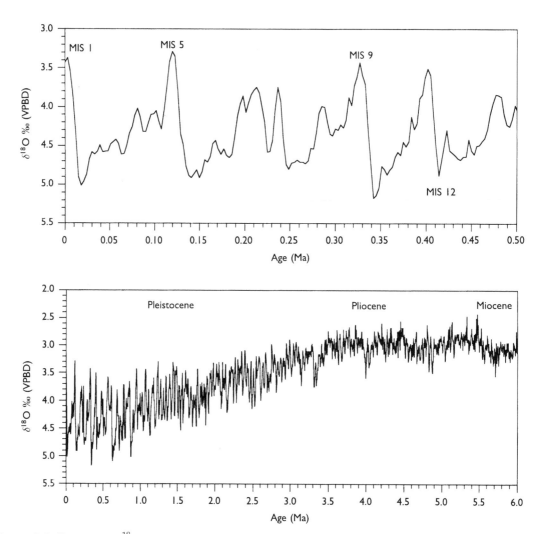

FIGURE 5.2 Composite $\delta^{18}O$ record for the last 500,000 years (upper panel) and the last 6 million years. The curves are based on Shackleton and Pisias (1985); Shackleton *et al.* (1990, 1995a,b)

Box 5.1 Major glaciations and interglacials of Europe and North America

Northern Europe		The Alps		North America		Great Britain		Russia	
Glacial		*Glacial*		*Glacial*		*Glacial*		*Glacial*	
	Interglacial		*Interglacial*		*Interglacial*		*Interglacial*		*Interglacial*
Weichsel		Würm		Wisconsin		Devensian		Valdai	
	Eem		Riss/Würm		Sangamon		Ipswichian		Mikulino
Saale		Riss		Illinoian		Wolstonian		Middle Russian	
	Holstein		Mindel/Riss		Yarmouth		Hoxnian		Likhvin
Elster		Mindel		Kansan		Anglian		White Russian	
	Cromer		Günz/Mindel		Aftonian		Cromer		Morozov
Menap		Günz		Nebraskan		Beestonian		Odessa	

5.2.2 The marine record

Evidence of changes in the nature and distribution of ocean water masses has been obtained from planktonic and benthic foraminifera, and from microfloral remains in deep ocean cores. Together with lithostratigraphic variations (for example ice-rafted debris), these data sources provide records of oceanographic changes (e.g. Broecker *et al.*, 1985). The global marine oxygen isotope record for the last 2.5 million years (Fig. 5.2) shows numerous oscillations mainly reflecting global ice volume changes. Before the late Pleistocene, glacial fluctuations had a dominant periodicity of about 40,000 years. Distinct cycles of 100,000 years duration are only present for the last 700,000 years. Interglacials as warm as the present (Holocene) occurred for only about 10 per cent of the Quaternary.

High-resolution marine records show that a number of distinctive, high-frequency events occurred during the last interglacial/glacial transition at 115,000–70,000 yr BP (Fig. 5.3). The oxygen isotope records from deep ocean cores show two excursions towards colder climates and more extensive glaciation (stages 4 and 2) with an intervening period (stage 3) of isotopically lighter values. The oxygen isotope stage 4/3 and 3/2 transitions are dated by means of orbital tuning to 58,000 and 23,000 yr BP, respectively. The marine record indicates that sea level rose from below −75 m to about −50 m between oxygen isotope stages 4 and 3, followed by a fall in sea level to approximately −120 m during isotope stage 2. During these three isotope stages there are, however, at least six isotope excursions between 60,000 and 20,000 yr BP. Cores from the North Atlantic show, as an example, evidence of repeated southern shifts of polar waters during the last 100,000 years, correlating with high-frequency climatic signals obtained from Greenland ice cores (Bond *et al.*, 1993; McManus *et al.*, 1994). These rapid climatic oscillations appear to have occurred simultaneously with cold meltwater influxes from the Laurentide and Fennoscandian ice sheets. Large meltwater discharge periods took place around 15,000–14,500, 13,500,

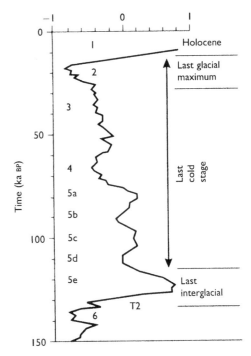

FIGURE 5.3 Oxygen isotope curve through isotope stages 1–5 (Weichselian and Eemian). (Adapted from Martinson *et al.*, 1987)

12,000 and 10,500 yr BP. The most significant of these was the Younger Dryas cooling (11,000–10,000 yr BP) when surface waters in the North Atlantic may have been chilled by the diversion of Laurentide ice-sheet meltwater from the Mississippi to the St Lawrence drainage routes (e.g. Broecker *et al.*, 1989). Rapid changes in the flow of warm Atlantic surface water involved shifts in sea-surface temperatures of about 5°C or more in less than 40 years (Lehman and Keigwin, 1992). Movement of ocean water masses can also be explained by internal factors such as circulation changes caused by salinity (density) and water temperature gradients. This process is called the *thermohaline circulation*. In the North Atlantic, this process operates like a conveyor belt; water flows northwards in the surface layers, to sink at around 60°N to form a deep-water mass termed the *North Atlantic Deep-Water* (NADW). It has been proposed that differences in salt concentration between the Atlantic and Pacific Oceans drive a

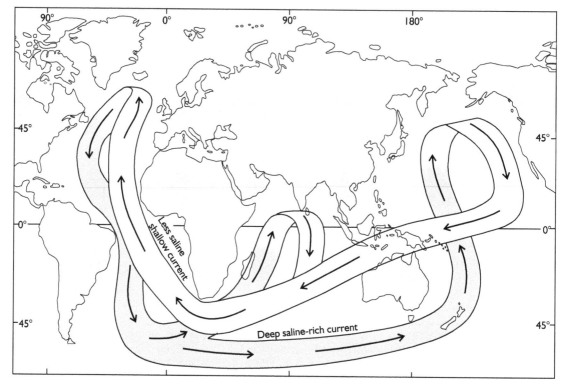

FIGURE 5.4 The large-scale salt transport system operating as a global ocean conveyor belt. (Adapted from Broecker and Denton, 1990)

Global Conveyor (Broecker and Denton, 1990) (Fig. 5.4).

In the North Atlantic, the climatic impact of the NADW formation is significant. During winter, ventilation of the ocean by sinking of the surface water releases about 8 calories m^{-3} of water. The annual heat released to the atmosphere from this process is equivalent to about 25 per cent of the solar heating at the sea surface north of 35°N (Broecker and Denton, 1990). A switch-on/off mechanism for the North Atlantic conveyor has been postulated to explain rapid changes in ocean circulation in the North Atlantic during the last glacial/interglacial cycle (e.g. Broecker *et al.*, 1990). Changes in the production and circulation of deep-water can be reconstructed from changes in cadmium and carbon isotopes in benthic foraminifera, since they reflect changes in nutrient budgets in connection with deep-water changes (e.g. Boyle, 1988).

5.3 Late Cenozoic glacier and climate variations

5.3.1 The Arctic

The PONAM (Polar North Atlantic Margin: Late Cenozoic Evolution) Programme (1989–1994) has given new insight into the major climate-driven environmental variations in the Norwegian–Greenland–Nordic Seas and their adjacent continental margins over the past 5 million years, and during the last glacial–interglacial cycle in particular (Elverhøi *et al.*, 1998). The Svalbard–Barents Sea ice sheet has been highly dynamic; after reaching the shelf edge during each stadial, it almost disappeared during interstadials. The east Greenland ice sheet, by contrast, showed only small advances in the fjords and inner shelf areas. The first Weichselian glacial advance of the two ice sheets in east Greenland and on

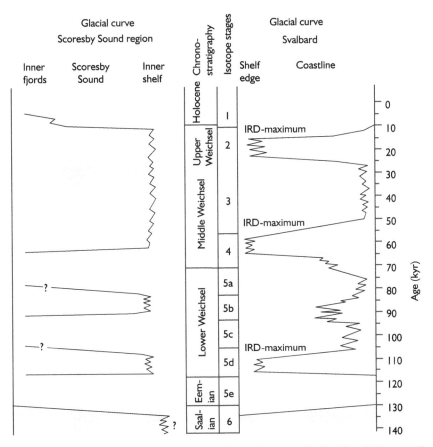

FIGURE 5.5 Time–distance diagram for Weichselian glaciations on the Svalbard–Barents Sea continental margin and the Scoresby Sound region (East Greenland margin). (Modified from Elverhøi *et al.*, 1998)

Svalbard occurred simultaneously (Fig. 5.5). Subsequent to 65,000 yr BP, however, the chronology and dynamics of the two ice sheets were different. The Svalbard–Barents Sea ice sheet experienced significant Middle and Late Weichselian ice advances, whereas the east Greenland ice sheet was characterized by a 55,000 yr period with a stable ice margin in the fjords or inner shelf areas (Elverhøi *et al.*, 1998). The different behaviour of the two ice sheets was probably caused by the palaeoceanographic circulation pattern in the polar North Atlantic. The Svalbard–Barents Sea ice sheet was under the influence of the relatively warm North Atlantic Current, whereas the east Greenland ice sheet was affected by the cold East Greenland Current. The east Greenland ice sheet seems to have developed during the Middle/Late Miocene, while the Svalbard–Barents Sea ice sheet developed during the Late Pliocene. The margin of the Svalbard–Barents Sea was characterized by major prograding fans formed by a series of debris flows, interpreted as products of rapid sediment delivery from fast-flowing ice streams which reached the shelf break during full glacial conditions. Off east Greenland, such submarine fans are not found north of Scoresby Sound. The east Greenland ice sheet rarely reached the shelf break, the sedimentation rates were relatively low, and the sediment transport was localized to several major deep-sea submarine channel systems (Elverhøi *et al.*, 1998).

The glaciation history of Svalbard (78°N) and the NW Barents Sea during the last 130,000 yr, as reconstructed by Mangerud and Svendsen (1992) (see major review by Mangerud *et al.*, 1998) from coastal cliff sections at the head of Isfjorden, shows four separate tills intercalated by marine sediments. The lowest marine formation contains *Mytilus edulis*, reflecting higher temperatures than at present, and correlates with the Eemian interglacial (deep sea oxygen isotope stage 5e). The three tills above reflect major glaciations around 110,000 yr BP, 75,000–50,000 yr BP, and 25,000–10,000 yr BP. The ice-free periods occurred between 110,000–75,000 yr BP (the Phantomodden interstadial) and between 50,000 and 25,000 yr BP (the Kapp Ekholm interstadial). Marine faunal assemblages indicate seasonally ice-free conditions. The reconstructed glacial history during the last interglacial/glacial cycle led Mangerud and Svendsen (1992) to postulate that the Quaternary glaciations on Svalbard have been driven by the 41,000 yr tilt period, in contrast to the Scandinavian glaciations primarily driven by the 21,000 yr precession cycle.

Several authors have claimed that the west coast of Spitzbergen remained ice-free during the Late Weichselian (e.g. Miller *et al.*, 1989; Forman, 1989). Mangerud *et al.* (1992) claimed, on the other hand that the Late Weichselian glaciation was more extensive than previously suggested, based on evidence from till-covered terraces, sea-level studies and radiocarbon dates. Studies of the Late Weichselian glacial history of the continental shelf off western Spitzbergen show that the outer part of Isfjorden and the inner shelf of this fjord are dominated by a relatively thin layer (10–20 m) of glacigenic sediments without any ice-marginal deposits (Svendsen *et al.*, 1992). The sediment thickness, however, increases towards the outer shelf, reaching more than 500 m at the edge of the continental shelf. In contrast to the inner shelf, possible moraine complexes were identified on the outer shelf. Radiocarbon dates on shells show that the basal glaciomarine deposits resting on till were deposited around 12,500 yr BP. Studies of the lacustrine sediment sequence in

Linnévatnet indicate rapid deglaciation without any Late-glacial readvances after 12,500 yr BP (Mangerud and Svendsen, 1990). During the Younger Dryas, glaciers in the Linnévatnet catchment were even smaller than during the Little Ice Age, in contrast to the situation in NW Europe (Svendsen and Mangerud, 1992).

From Edgeøya and Barentsøya, east of Svalbard, four relative sea-level curves have been constructed (Bondevik *et al.*, 1995). Subsequent to the formation of the marine limit at about 11,500 cal yr BP (ca. 10,000 ^{14}C yr BP), the rate of land uplift was of the order of approximately 40 mm yr^{-1}. A minimum rate of uplift at 8000 cal yr BP was explained by decreased isostatic uplift in combination with eustatic sea-level rise. A steady emergence has taken place over the last 7000 radiocarbon yr BP. Bondevik *et al.* (1995) suggested that the regional Holocene uplift pattern reflects a larger glacio-isostatic depression in the southern Barents Sea during the last glaciation.

For the past two decades, the glacier extent during the last glacial maximum over Queen Elizabeth Island (see England, 1998, and references therein) and in the NW Territories of Canada (e.g. Dyke *et al.*, 1982) has been debated. The maximum model indicates a regional Innuitan ice sheet which coalesced with the Greenland ice sheet to the east and with the Laurentide ice sheet to the south. An alternative model suggests a non-contiguous array of alpine ice caps, termed the Franklin Ice Complex. Evidence presented from the eastern side of the Eureka Sound and elsewhere in the region (e.g. England, 1998) strongly supports the existence of an Innuitan ice sheet with glaciers covering the Nares Strait during the last glacial maximum. On Ellesmere Island, abundant ice-marginal features lie in a 500 km long zone with a distal margin 10–60 km beyond the present ice caps (Hodgson, 1985). At the head of the fjords, this drift belt was deposited between 9000 and 7000 yr BP. Subsequently the glaciers retreated, followed by a late Holocene glacier readvance (England, 1986). Despite the fact that the ELA was at about 500 m during the last glaciation, the glacier extent was rather limited (normally 5–40 km beyond present ice margins). The area must therefore have been characterized by extreme continentality (cold summers and less precipitation than at present). Recent dating evidence obtained from cosmogenic isotopes presented by Zreda *et al.* (1999) also suggests that the Nares Strait was filled with ice during the last glaciation, blocking the connection between the Arctic and Atlantic oceans, thereby supporting the model of extensive and long-lasting terrestrial glaciers in this region.

5.3.2 Eurasia

The research programme 'Quaternary Environments of the Eurasian North' (QUEEN) has recently provided new information on the last glacial/interglacial cycle in northern Eurasia (Thiede and Bauch, 1999). Eemian marine and estuarine sediments, identified by their warm boreal benthic faunas, are exposed along the large rivers of northern Russia between the Kola Peninsula in the west and Taymyr in the east (Larsen *et al.*, 1999a; Mangerud *et al.*, 1999). Several contrasting models have been proposed for the last Eurasian ice sheet over Russia. Grosswald (1980, 1993, 1998) and Grosswald and Hughes (1995) proposed that a large ice sheet had existed over this region. Velichko *et al.* (1997), on the other hand, suggested an intermediate-sized Valdaian (Weichselian) glacial maximum. The northernmost (and youngest) ice-marginal zone between the White Sea and the Ural mountains is the W–E-trending belt of a 700-km-long moraine system, termed the Markhida Line (Fig. 5.6). Along the Markhida Line, it has been shown that the morphological expression of ice-marginal features becomes progressively fresher from west to east, although evidence

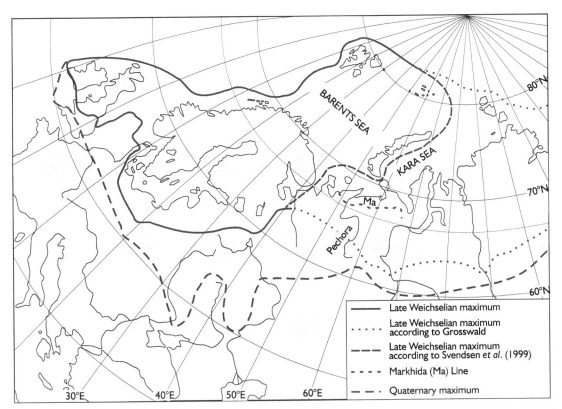

FIGURE 5.6 Map of Eurasia with glacial limits. The glacial limits over northern Russia and in the Kara/Barents seas are adapted from Mangerud *et al.* (1999) and Svendsen *et al.* (1999)

suggests that the moraine is of the same age. This has been attributed to permafrost and buried glacier ice having survived much longer in the east than in the west (Astakhov *et al.*, 1999). Reconstruction of the surface profile of the ice sheet reaching the Markhida Line suggests a very low gradient surface profile over a thick sedimentary substrate, while the surface profile was steeper near the Urals (Tveranger *et al.*, 1999).

Yakovlev (1956) proposed a restricted glacial maximum, with its southern limit following the Markhida moraine. Glacial elements indicate that this moraine system was deposited by the Kara ice sheet, and in the west, by the Barents ice sheet. The moraine is stratigraphically above Mikulino (Eemian) marine sediments (Mangerud *et al.*, 1999). Beyond the moraines are Eemian marine sediments and three archaeological sites (not overrun by glaciers) with radiocarbon dates in the range of 37,000–16,000 yr BP. Grosswald's Late Valdaian maximum ice limit is 400–700 km further to the south. Based on a series of radiocarbon and OSL dates of >45,000 yr BP proximal to the Markhida line, Mangerud *et al.* (1999) concluded that the Markhida moraine is of Middle/Early Valdaian age. Terrestrial and marine evidence led Svendsen *et al.* (1999) to suggest that the Late Valdaian/Weichselian ice margin of the Kara Sea ice sheet did not reach the present coast of northern Russia (Fig. 5.6), indicating that the Late Valdaian/Weichselian glacial maximum was less than half of the maximum model which has been used as a boundary condition in global circulation models (GCMs).

It has been demonstrated that glaciation on the Kara Sea shelf (Svendsen *et al.*, 1999), in western Siberia (Astakhov, 1992), in the Pechora Basin (Astakhov *et al.*, 1999; Mangerud *et al.*, 1999) and on Taymyr (Möller *et al.*, 1999) culminated before the Late Valdaian, out of phase with the global ice volume maximum at around 20,000 yr BP. Glacial terrestrial sediments dated to the Late Valdaian have been identified along the Severnaya Dvina, where a tongue from the Fennoscandian ice sheet moved southeastward in the wide river basin, with a maximum ice extent subsequent to 17,000 yr BP

(Larsen *et al.*, 1999b). Earlier syntheses of the development of the Scandinavian, Barents Sea and Kara ice sheets have generally postulated that they fluctuated in phase (e.g. Denton and Hughes, 1981). According to the model presented by Svendsen *et al.* (1999), this concept must be changed. During the Early–Middle Valdaian/Weichselian when the Kara Sea ice sheet in the east reached its maximum extent, the Scandinavian ice sheet in the west was less extensive than at its Late Valdaian/Weichselian maximum stage. In the Late Valdaian/Weichselian, in contrast, when the Scandinavian ice sheet culminated, the Kara Sea ice sheet was small or absent. The role of the Barents Sea ice sheet is, however, still unresolved. At its western margin over Svalbard, two (Middle and Late) Valdaian/Weichselian maxima occurred (Landvik *et al.*, 1998; Mangerud *et al.*, 1998).

According to Grosswald, the melting of the ice sheet was dominated by a bipartition of the meltwater along the SW margin, with an initial phase of drainage towards the south to the Caspian Sea and the Black Sea, and later towards the west to Poland and Germany. In the final deglacial phase, the meltwater drainage was northwards to the Barents Sea. Based on studies of raised Holocene shorelines in northern Novaya Zemlya, Forman *et al.* (1999) suggested that the Barents Sea ice sheet either melted at approximately 13,000 yr BP or was very thin.

The palaeoclimatic record from the bottom sediments of Lake Baikal, East Siberia, has revealed evidence for an abrupt and intense glaciation during the initial part of the last interglacial period (isotope substage 5d). The glaciation lasted about 12,000 years from 117 to 105 kyr BP, according to correlation with the SPECMAP isotope chronology (Karabanov *et al.*, 1998). The authors suggested that this glaciation was caused by cooling due to an orbitally-driven decrease in solar insolation, coupled with western atmospheric transport of moisture from the open areas of the North Atlantic and Arctic seas.

On the Tibetan Plateau, several terminal moraines sequences have been studied extensively; however, a homogeneous chronology is still lacking. So far, Late-glacial terminal moraines have been found in the western

Kunlun Shan and in the Tian Shan. Radio-carbon dates of around 15,000 yr BP correspond to a glacier advance south of the Qaidam Basin. Except for glacier readvances at about 3000 yr BP and during the Little Ice Age, the timing of glacier advances during the last glaciation on the Tibetan Plateau are in fact unknown (Lehmkuhl, 1997). There has been considerable debate concerning the glacier extent on the Tibetan Plateau at the last glacial maximum (LGM) (e.g. Shi *et al.*, 1992). Kuhle (1987, 1988, 1991, 1998) postulated an ice sheet with a thickness of 2.5 km and an areal extent of 2–2.4 million km^2. This means that, according to his reconstruction, an ice cap covered almost the entire Tibetan Plateau. An opposite view has been proposed by Chinese investigators, who concluded that glaciers were restricted to the main mountain areas with outlet glaciers discharging on to the plateau (Zheng, 1989). The areal extents of LGM and older glaciations discussed by Chinese investigators (for details, see Lehmkuhl, 1997; Lehmkuhl *et al.*, 1998) differ from each other, but they all show only restricted mountain and plateau glaciation. In the eastern and central parts of the Tibetan Plateau, the LGM glaciation seems to have been restricted to isolated mountain massifs (e.g. Lehmkuhl, 1994), in agreement with the investigations by the Chinese researchers. At the northern margin of the Tibetan Plateau, however, the LGM glaciers were probably more extensive than proposed by the Chinese researchers (e.g. Hövermann and Lehmkuhl, 1993).

Minimal and maximal models of Late Pleistocene glaciation on the Tibetan Plateau were evaluated by Kaufmann and Lambeck (1997) using present uplift rates, free-air gravity anomalies and eustatic sea-level rise. Their results are more in support of the large ice sheet proposed by Kuhle *et al.* (1989) than the more restricted glaciation proposed by Gupta *et al.* (1992).

In the NW Himalayas and Karakoram Mountains of Pakistan and in NW India, three glacial events, progressively less extensive with time, have been mapped in most regions (Derbyshire and Owen, 1997; Owen *et al.*, 1998). The ages of these glacial stages

are, however, poorly known due to lack of the dating material necessary to establish an absolute chronological framework. In the Hunza Karakoram, glacier fluctuations at the end of the Pleistocene seem to have been characterized by several halts or readvances.

In the Mongolian mountains, three areas bear evidence of Quaternary glaciations; Khentey, Khangay and Mongol Altai (Lehmkuhl, 1998). The glacial landforms there consist of mainly cirques, U-shaped valleys and hanging valleys. Based on weathering of sediments and morphostratigraphy, two major Pleistocene glaciations have been distinguished.

The evidence of late Pleistocene glaciation in the southern ranges of the former Soviet Union is fragmentary and indirect (Bondarev *et al.*, 1997). Up to the early Holocene, no moraines have been dated securely. A valley-type model (in contrast to an ice sheet model) of Pleistocene glaciation has been proposed. Up to five glacier advances have been tentatively reconstructed for the Late Pleistocene. Most areas were deglaciated by 10,000 yr BP, after which only small valley glaciers readvanced.

5.3.3 Northern Europe

At its maximum, the Eurasian ice sheet extended eastwards to the Ural Mountains, southeast beyond Kiev, southwards into central Germany and westwards to the British Isles. The subdivision into several ice ages is based on morphostratigraphic evidence, with progressively younger terminal moraine systems northwards. The different stages were named (from the oldest) Elster, Saale and Weichsel. The Saale glaciation was later subdivided into the Drenthe and Warthe moraine stages. All stages are considered to have been deposited during the last 900,000 years (e.g. Sibrava, 1986). Deposits from the Elster stage represent the earliest definite evidence of major glaciation in northwestern Germany and Europe. These deposits occur in connection with a series of deep, buried channels formed by subglacial meltwater. In contrast with the low- or no-relief deposits from the Elster glaciation, deposits from the Saalian and especially from the Weichselian glaciation exhibit distinct morphological features. The interglacial stages are represented by deposits of marine transgressions in the lower areas, and by terrestrial peat with pollen reflecting the vegetation during the different interglacials.

The glacial stages in Britain equivalent to those on mainland Europe are considered to be the Anglian, Wolstonian and Devensian (West *et al.*, 1988). In the British Isles, several glaciation centres existed during the Quaternary, and combined with frequent phases of coalescing with the Fennoscandian ice sheet, the moraine sequence is less evident there than in mainland Europe.

In northern Europe, the record of glacier variations during the Late Cenozoic has been reconstructed from terrestrial data and offshore ice-rafted debris (IRD) from dated sediment cores from the Norwegian Sea (Fig. 5.7). The input of IRD is used as a proxy for ice sheet advances reaching the shelf (e.g. Jansen and Sjøholm, 1991; Baumann *et al.*, 1995). The first glaciation to reach the coast of the Nordic Seas (Scandinavia, Svalbard and/or Greenland) occurred at about 11 million yr BP (Late Miocene), as recorded by IRD in the Norwegian Sea (Mangerud *et al.*, 1996). Oxygen isotope records, IRD curves and palynological evidence

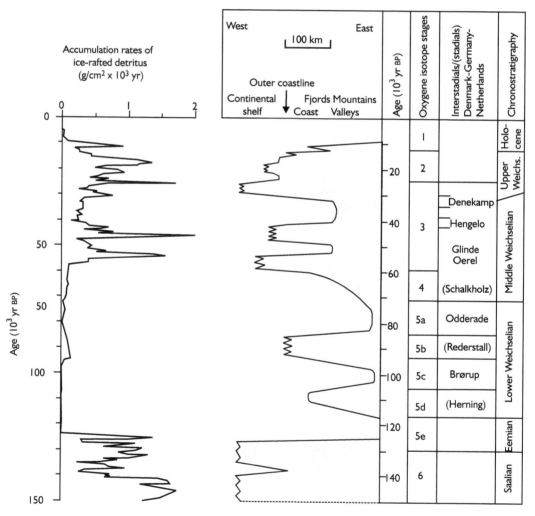

FIGURE 5.7 Glaciation curve of western Scandinavia based on terrestrial (right graph) and marine data (left graph). (Adapted from Baumann *et al.*, 1995)

from the Netherlands indicate that the major glaciations in Scandinavia and Svalbard started around 2.8 million yr BP. During the last 2.6 million years, warm interglacials like the Holocene did not last for more than 6–8 per cent of the time, whereas glacial maxima, like the 20,000–18,000 yr BP Late Weichselian maximum, occupied less than 5 per cent of the time. The large-amplitude climate and ice-sheet fluctuations have occurred mainly during the last 900,000 years (Mangerud et al., 1996).

During the Saalian glaciation the western margin of the Scandinavian ice sheet advanced on to the shelf, probably reaching the edge of the continental shelf. At the Saalian/Eemian transition, the ice sheet retreated rapidly and decayed perhaps within 2000 years or less. The Eemian, as initially defined in the Netherlands and correlated all over northern Europe, correlates with oxygen isotope stage 5e (Mangerud, 1989). During the Early Weichselian (oxygen isotope substages 5d–5a), two glacier advances occurred. The first (during substage 5d) did not reach the coast, whereas the second (5b) advanced to the coastline and released icebergs in some areas. During the Early Weichselian, two interstadials occurred in northwestern Europe, the Brørup/ St. Germain I and Odderade/St. Germain II (in Scandinavia represented by the Jämtland/ Peräpohjola and Tärendö interstadials, respectively), which have been correlated with oxygen isotope stages 5c and 5a, respectively (Andersen and Mangerud, 1989; Mangerud, 1991). In general, oxygen isotope stage 5 in western Scandinavia was characterized by low IRD deposition and short glaciation phases (Baumann et al., 1995). In the Middle Weichselian, glaciers may have begun to advance during the early part of oxygen isotope stage 4 (ca. 70,000 yr BP). According to the IRD signal, however, the ice sheet did not reach its maximum position before ca. 63,000 yr BP. This glaciation phase was terminated by a deglaciation phase at approximately 54,000 yr BP. A glaciation dated to 47,000–43,000 yr BP on the terrestrial record is also found in the IRD record. Some thermoluminescence (TL) and U-series dates indicate that central southern Norway was deglaciated at 40,000–30,000 yr BP. A significant

deglaciation phase is recorded both in the marine and terrestrial record between 38,500 and 32,500 yr BP (the Denekamp/Ålesund interstadial) (Andersen and Mangerud, 1989). During the Late Weichselian, several ice-sheet oscillations occurred; the highest mountains in Scandinavia most probably stood above the ice sheet as nunataks (Nesje et al., 1988; Nesje and Dahl, 1992).

The earliest Late-glacial warming in Europe occurred around 15,000 yr BP (Artemesia expansion). In some areas there is evidence of warming from around 13,500 yr BP; however, the most significant warming is recorded in pollen and Coleoptera assemblages from about 13,000 yr BP (Lowe and Walker, 1997). The thermal maximum of the Late-glacial interstadial complex occurred at 13,000–12,500 yr BP in Britain, The Netherlands, southwest Europe and Switzerland, between 12,500 and 12,000 yr BP in southern Scandinavia and Germany, and between 11,500 and 11,000 yr BP in SW and northern Norway (Fig. 5.8). Marked climatic gradients during this period most probably reflect the cooling effects of the retreating Scandinavian ice sheet and changing thermohaline circulation in the North Atlantic.

In the Late-glacial sequence in northern Europe, a series of distinct climatic oscillations occurred, mainly recorded in pollen sequences. Bølling was a mild interstadial 13,000–12,000 yr BP, followed by the cool and short Older Dryas stadial 12,000–11,800 yr BP. Allerød was a mild interstadial between 11,800 and 11,000 yr BP, followed by the significant Younger Dryas cooling and glacier expansion between about 11,000 and 10,000 yr BP (Mangerud et al., 1974). This cooling caused readvances of the Scandinavian ice sheet and expansion/reformation of cirque glaciers beyond the continental ice sheet, especially along the western margin. In Scotland, a 2000 km^2 ice field developed (e.g. Sissons, 1979b; Thorp, 1986; Ballantyne, 1989), whereas minor valley and cirque glaciers formed in the upland areas in Scotland, England, Wales and Ireland (Gray and Coxon, 1991).

Biostratigraphical sequences from Fennoscandian sections were correlated by Forsström and Punkari (1997) with reference sequences

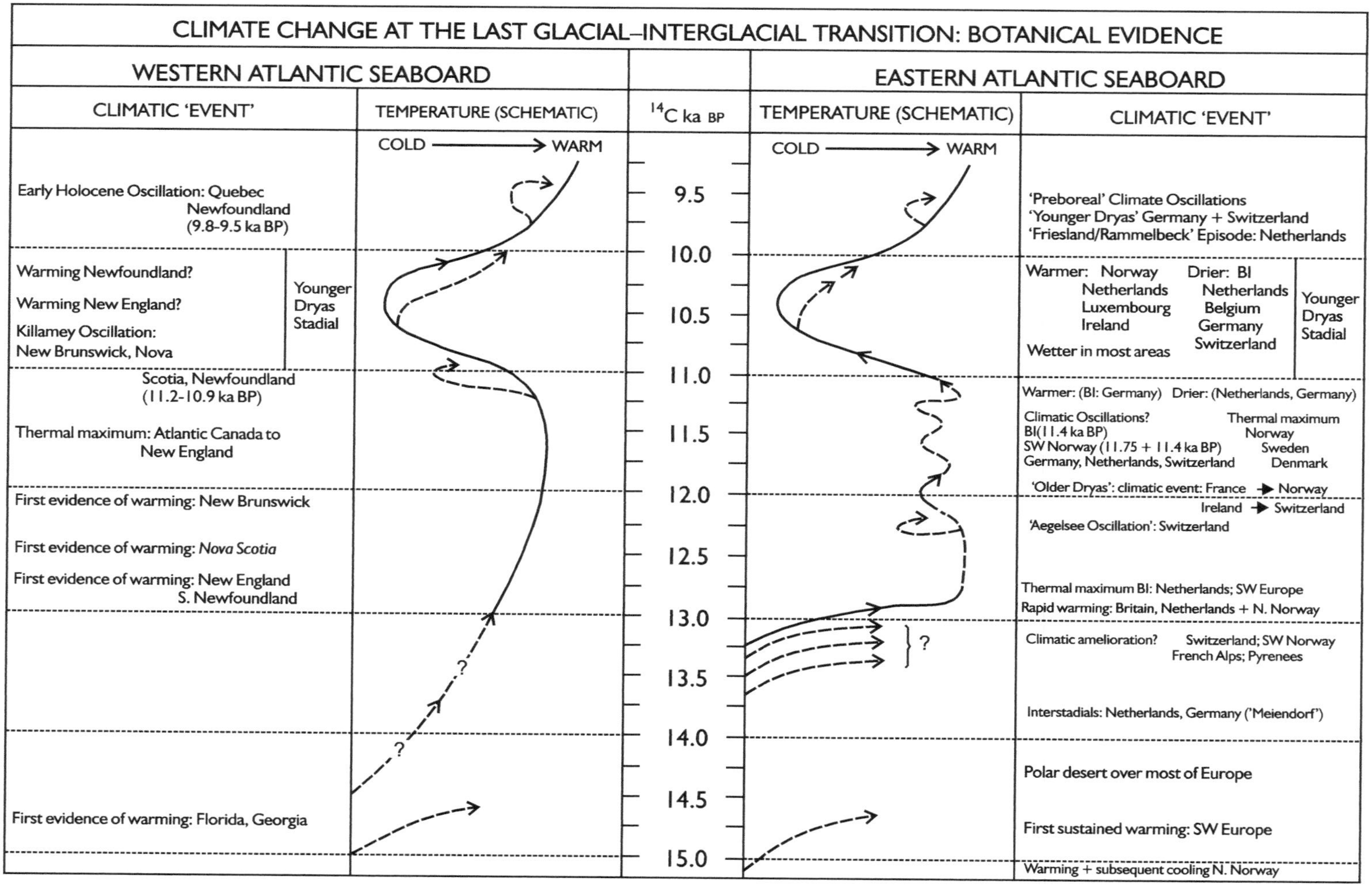

Figure 5.8 Climatic changes in Europe and eastern North America during the last glacial–interglacial transition. (Modified from Lowe and Walker, 1997)

from Estonia and with sections located near or beyond the margins of the last glaciation. Organic sediments previously attributed to Early and Middle Weichselian interstadial periods in Finland were argued by them to be redeposited and mixed older material from the last interglacial (the Eemian). They suggested that the Eemian climatic optimum was followed by a continuously cooling climate and a marine regression. Their reinterpretation suggests that the ice sheet grew over Finland during the first Early Weichselian stadial. The preservation of the interglacial beds and the lack of younger non-glacial sediments, they argued, support the interpretation that the area remained ice-covered until the final deglaciation.

During the last glacial–interglacial transition, the movements of the North Atlantic Polar Front have been described as hinging around locations in the western North Atlantic. Iceland, situated in the middle of the North Atlantic Ocean, has glaciers sensitive to changes in the oceanic and atmospheric front systems (Ingolfsson *et al.*, 1997). The Late-glacial records from Iceland indicate that relatively warm Atlantic water reached Iceland during the Bølling–Allerød interstadial complex, with a short cooling period corresponding to the Older Dryas. Karpuz *et al.* (1993) suggested that the marine polar front was located close to Iceland during Bølling–Allerød, and Sarnthein *et al.* (1995) concluded that sea-surface circulation was mainly in a Holocene interglacial mode after 12,800 yr BP. Like elsewhere in NW Europe, an abrupt cooling marks the beginning of the Younger Dryas. Terrestrial data from Iceland demonstrate a transition from mild climatic conditions by the end of the Allerød, to polar conditions and significant glacier expansion. Pollen influx dropped significantly and the content of organic carbon in lake sediments from northern Iceland demonstrates rapid climatic change. The sequence of deglaciation and terrestrial biostratigraphical records indicate that the Preboreal was a period of ameliorating climate (Rundgren, 1995), and by ca. 8000 yr BP, glaciers were of similar size as at present (Kaldal and Vikingsson, 1991).

Ash Zone 1 in Iceland consists of at least five different tephra populations deposited over a period of ca. 1500 radiocarbon years, as recognized in lake sediments from Skagi, northern Iceland (Björck *et al.*, 1992), of which the Vedde and the Saksunarvatn ash layers are the most widely recognized.

5.3.4 The Alps

At their maximum, Alpine glaciers covered about 150,000 km². The Alpine glaciers flowed as a network among mountain peaks and ice divides, with coalescing valley and piedmont glaciers. Based on work in the northward-draining valleys of the German Alpine foreland, south of Munich, Penck and Brückner (1909) presented a scheme of glacials (Würm, Riss, Mindel, Günz) and interglacials (Würm/Riss, Riss/Mindel, Mindel/Günz) (Box 5.1). Subsequently, the four-part sequence was extended by the discovery of two older glaciations, the Donau and Biber. The four main glacial stages are represented by a series of terraced glaciofluvial outwash plains ('schötter'); each younger and, in general, lower plain was related to terminal moraines further upvalley. During the interglacials, these were, according to Penck and Brückner (1909), eroded to form terraces. Problems with this classic Alpine sequence have been outlined (e.g. Sibrava, 1986). The sequence of terraces is more complicated than originally proposed, and contains both interglacial and postglacial material. In addition, the erosion is rather glacial than interglacial, and the deposits may represent only a few millennia of glaciation (Kukla, 1977). The classical nomenclature therefore only has morphostratigraphic significance in the study area of Penck and Brückner, and must be abandoned for external correlations (Sibrava, 1986).

The glacial history of the eastern European Alps during the LGM has been reconstructed by mapping, palynology and radiocarbon dating. During the glacier build-up toward the LGM, topographical constraints in the form of deep valleys led to glaciers occupying tributary valleys and troughs until about 24,000 yr BP (van Husen, 1997). Subsequently, rapid glacier expansion in the main valleys led to ice streams and piedmont glaciers in the Alpine foreland. Radiocarbon dates obtained from organic material in the outwash ('Niederterrasse') show that the build-up ended around 21,000 yr BP. According to the outwash deposits, the LGM lasted for about 3000–4000 years. The deglaciation from the LGM was apparently very rapid. The glacier retreat was interrupted by minor oscillations at around 16,000 yr BP (Oldest Dryas), 14,000 yr BP (Gschnitz Phase), during the Older Dryas at ca. 12,000 yr BP (Daun Phase), and finally during the Younger Dryas between 11,000 and 10,000 yr BP (Egesen Phase).

Glacial evidence in the Gran Sasso Massif of the central Apennines in Italy has led to the reconstruction and dating of the last glacial maximum advance and subsequent readvance phases (Giraudi and Frezzotti, 1997). During the Campo Imperatore Stade (22,600 yr BP) glaciers reached their maximum extent. During this phase, mean annual temperatures were of the order of 7–8°C lower than at present, and the amount of snowfall was similar to present. The glaciers started to retreat approximately 21,000 yr BP, forming three recessional moraines between 21,000 and 16,000 yr BP. Glacier retreat subsequent to 15,000 yr BP left behind another four moraines.

There has been a growing recognition that the Egesen moraines in the Alps were deposited during the Younger Dryas (e.g. Kerschner *et al.*, 1998). Surface exposure dates of Egesen moraines in Julier Pass, Switzerland, showed that the moraines were deposited during the early part of the Younger Dryas chronozone. In some valleys, numerous Egesen moraines are present, indicating oscillating glaciers. The moraine complex has been divided into three or, in places, four distinct groups. Snowlines, treelines and rock glaciers have been used to calculate temperature depressions and precipitation changes for the Younger Dryas using glacial–meteorological and statistical models. From these calculations, summer temperatures may have been about 3°C lower than at present, while annual temperatures were at least 4–6°C lower in the central Alps. Precipitation during the Younger Dryas was probably about the same as at present in the northern and western parts of the Alps, and decreased significantly towards the interior and the south. At the end

of the Younger Dryas, precipitation decreased and in the Central Alps of Austria and eastern Switzerland, the climate was almost semi-arid (Kerschner *et al.*, 1998).

5.3.5 North America

The Laurentide ice sheet extended from the Arctic Ocean in the Canadian Arctic archipelago to the mid-western states in the south, and from the Canadian Rocky Mountains in the west. The most extensive record of fluctuations along its southern margin comes from the north central United States. Named after the states where they are best characterized, the Nebraskan, Kansan, Illinoian and Wisconsin glaciations represent the glacial sequence (Box 5.1), the Kansan considered to be the most extensive glaciation. The earliest three glaciations are based on till sheets, while the Wisconsin was based on terminal moraines. The interglacials were based on palaeosols developed in tills. Based on new evidence provided by means of new methods and extensive fieldwork, the original stratigraphic nomenclature has been challenged (e.g. Hallberg, 1986).

During periods of maximum Quaternary glaciation, including the Wisconsin glaciation, the continental ice sheet was more or less continuous over the North American continent. The ice sheet consisted of two main parts: the Cordillera ice sheet, centred in the Coastal range and Rocky Mountains in the west, and the Laurentide ice sheet in the east. The former was most extensive in the British Colombian mountains. The southern limit for continuous ice was at the Columbia River south of the Canada/USA border. The Laurentide ice sheet was, together with the Eurasian ice sheet, responsible for most of the glacio-eustatic lowering of sea-level of ca. 120 m during the LGM. Inferred from the pattern of postglacial uplift, the ice was thickest over Hudson Bay. The different parts of the Laurentide ice sheet reached their maximum extent between 22,000 and 17,000 yr BP. The Cordillera ice sheet, however, reached its maximum extent approximately 15,000–14,000 yr BP. During its maximum extent, the Laurentide ice sheet was more than twice as big as the north European ice sheet. To the north, the ice sheet may have coalesced with ice over the Queen Elizabeth Islands. Morphological evidence suggests that the Laurentide ice sheet had two ice centres, one over Labrador and one over Keewatin.

In the Canadian and northern American Rockies, glacier fluctuations have been reconstructed using stratigraphy of glacial deposits, geomorphology, and lake and peat deposits. The history of glacier recession of the Late Wisconsin valley glaciers in the Canadian and northern American Rockies is not well documented (Osborn and Gerloff, 1997). Evidence presented so far suggests that glaciers retreated to within tens of kilometres of the present ice margins before ca. 12,000 yr BP. Moraines a few kilometres beyond Little Ice Age moraines indicate one or several readvances or stillstands. The Piper Lake moraine suggests a readvance before 11,200 yr BP. An age of ca. 11,000 yr BP from sediments above Late Wisconsin till at Crowfoot Lake indicates that glaciers had retreated to modern limits at that time. In the Mission Mountains, glaciers had retreated to within less than 1 km of the cirque headwalls by 11,200 yr BP. Data from Crowfoot Lake indicate that a minor readvance of Crowfoot Glacier occurred between ca. 11,300 and 10,000 yr BP (Osborn and Gerloff, 1997). Subsequent to the Crowfoot Advance, most glaciers in the Rockies retreated, as demonstrated by wood radiocarbon-dated at 8200 yr BP washed out from the base of the Athabasca Glacier. Osborn (1985) compared modern and Crowfoot ELAs using the median altitude approach. ELA depression during the Crowfoot advance ranged from about 5 m for small basin-filled glaciers, to 195 m for the large and steep Jackson Glacier, with a mean of 40 m ELA difference.

Based on organic content and magnetic susceptibility of continuous lake-sediment records of glaciations in Sierra Nevada, California, at least 20 stadial–interstadial oscillations between 52,600 and 14,000 yr BP are indicated (Benson et al., 1998). The record shows that a glaciation (Tioga) started at approximately 24,500 yr BP and terminated at around 13,600 yr BP. Alpine glacier oscillations in Sierra Nevada have occurred at a frequency of approximately every 1900 years during most of the last ~50,000 years. The Late-glacial Recess Peak advance in the Sierra Nevada was the first major glacier advance after retreat from the local Late Wisconsin (Tioga advance) glaciers. Dated lake cores suggest that the Sierra was deglaciated by 15,000–14,000 yr BP (Clark and Gillespie, 1997). Cirque moraines in Sierra Nevada show that the last significant pre-Little Ice Age advance (the Recess Peak of late Pleistocene age) resulted from ELA lowering of about twice that of the Little Ice Age (Matthes advance). Tephrochronology and radiocarbon dates from lacustrine sediments provide time constraints on the two advances. The absence of a young tephra on Matthes moraines in the central Sierra shows that they formed subsequent to 700 yr BP (ca. 650 cal years). The termination of the Recess Peak advance was established at 11,200 yr BP by extensive AMS radiocarbon dating on gyttja, peat, and macrofossils from cores. The evidence presented suggests that if there was an advance related to the Younger Dryas cooling, it was less extensive than the Matthes advance. In addition, the Matthes advance was the most extensive and most probably the only Neoglacial advance in the Sierra Nevada (Clark and Gillespie, 1997).

5.3.6 South America

In the eastern Colombian Andes, major glacial advances preceded the LGM at 18,000–20,000 yr BP (Helmens et al., 1997). Two glacial advances probably date between 43,000 and 38,000 yr BP and between 36,000 and 31,000 yr BP (isotope stage 3). These events are mapped from glacial sediments and moraine complexes. Three glacial advances are mapped between 23,500 and 19,500 yr BP, 18,000 and 15,500 yr BP, and the last one between 13,500 and 12,500 yr BP. The glacial landforms related to the 18,000–15,500 yr BP advance are the most distinct, forming arcuate, multiple ridges up to tens of metres high (Helmens et al., 1997).

Studies of late Quaternary moraines and lacustrine sediments around the southern Altiplano of Bolivia show a more-or-less synchronous development during the last glacial cycle. Radiocarbon dating of peat associated with glaciofluvial and glacial deposits indicate that the largest glacier advance at the end of the last glacial cycle culminated subsequent to 13,300 yr BP, and that a smaller glacier readvance possibly occurred between 12,000 and 10,000 yr BP. As the highest lake-level stand (Tauca phase) occurred at ca. 13,800 yr BP, it has been suggested that increased humidity and low temperatures during the time interval 14,000–13,000 yr BP were the main forcing factors for glacier expansion at that time (Clapperton *et al.*, 1997).

At the South Patagonian Icefield in Chile, the LGM margin was marked by moraines approximately 50 km from the present glaciers. At the icefield there is also evidence of two Late-glacial advances, represented by marginal deposits, including terminal moraines, 18–20 and 10–16 km from the margin of modern southern outlet glaciers from the icefield (Marden, 1997). Pumice clasts from an eruption of the Reclus volcano at about 11,900 yr BP gives a close limiting age for the older Late Glacial readvance, while the younger readvance occurred between 11,900 and 9200 yr BP.

A review of radiocarbon-dated glacier fluctuations in the northern and southern Andes indicated that seven to eight advances took place more or less synchronously during the interval from approximately 40,000 to 10,000 ^{14}C yr BP (Clapperton, 1998). The timings of these advances closely match the cold intervals identified in the records from the Greenland ice cores and North Atlantic deep-sea sediments, implying that there are mechanisms that can rapidly transmit a thermal signal through the atmosphere.

5.3.7 New Zealand

In the southern Alps of New Zealand, the LGM glaciers reached their maximum extent between 22,300 and 18,000 yr BP (Fitzsimons, 1997). A glacier readvance, almost reaching the extent of the LGM glaciers and possibly with two maxima, took place between 16,000 and 14,000 yr BP. Altogether, there is evidence for at least four advances between 18,000 and 8000 yr BP. Subsequent to 14,000 yr BP, the

glaciers retreated, however, interrupted by minor glacier readvances at around 11,000, 10,250 and 8600 yr BP (Fitzsimons, 1997). Although there is evidence of Late Pleistocene glaciation in the North Island of New Zealand, the advances are poorly dated and there is no evidence of glacier activity subsequent to the LGM (Shepherd, 1987).

5.3.8 Antarctica

The glaciation history of Antarctica between the end of the last interglacial (oxygen isotope stage 5e) and the last glacial maximum (LGM) (oxygen isotope stage 2) is not well known. It has been suggested that the West Antarctic ice sheet melted totally or partly during the last interglacial (Mercer, 1978; Denton and Hughes, 1981). It has also been speculated that high interglacial sea-levels may have caused rapid disintegration of the Ross and Weddell ice shelves, also influencing the dimensions of the ice caps since they were not buttressed by ice shelves. Some information, although controversial, is available about the extent of the ice sheet during the LGM. One opinion is that both the West and East Antarctic ice sheets extended to a position close to the edge of the continental shelf between 21,000 and 17,000 yr BP (Heusser, 1989). Another view is that the West Antarctic ice sheet was not much larger during the LGM than at present (see Denton and Hughes, 1981, for further discussion). Denton and Hughes (1981) suggested that the total ice volume of both the West and East Antarctic ice sheets during the last glacial maximum was 37 million km^3 (present volume 24 million km^3), mostly as a result of expansion of the East Antarctic ice sheet. This volume increase caused a 25 m lowering of global sea-level.

It has been suggested that the expansion of the West Antarctic ice sheet during the LGM resulted in the grounding of the Ross Sea and Weddel Sea ice shelf areas. Expansion of the mostly land-based East Antarctic ice sheet, by contrast, was restricted to 75 to 90 km because of the narrow surrounding continental shelf (Hollin, 1962). During the Late Quaternary, the prevailing opinion is that the Antarctic ice sheet fluctuated in phase with global sea-level changes, which again were influenced by the northern hemisphere Laurentide and Eurasian ice sheets (Hollin, 1962; Denton and Hughes, 1981; Labeyrie *et al.*, 1986; Heusser, 1989).

The development of the Antarctic ice sheet during the Late-glacial episode may, however, have been influenced by decreased precipitation or precipitation starvation due to sea-ice expansion (e.g. Mercer, 1983). Alpine glaciers on south Victoria Land retreated during the LGM, possibly due to this effect (Stuiver *et al.*, 1981; Mercer, 1983). Denton *et al.* (1971) proposed a more complex history for the Dry Valleys in south Victoria Land. During world-wide glaciation periods, with grounded ice in the McMurdo Sound and Ross Sea, glaciers advanced into the Dry Valleys, while adjacent land-based glaciers retreated due to reduced precipitation.

5.4 Late-glacial glacier and climate variations in NW Europe

High-frequency climatic fluctuations during the last deglaciation (ca. 14,000–9000 ^{14}C yr BP) are well documented in terrestrial data, marine records and ice cores from the North Atlantic region (e.g. Lowe and Walker, 1997, and references therein). These climatic changes occurred at a time of maximum solar radiation receipt in the northern hemisphere and cannot therefore be explained by orbital forcing. The cause(s) of these climatic fluctuations must therefore be sought in the ocean/atmosphere/climate system.

In northwestern Europe, several climatic oscillations occurred towards the end of the Weichselian/Devensian glaciation. These Late-glacial oscillations are dated to approximately 15,000–9000 radiocarbon years BP. Four periods are recognized based on biostratigraphical evidence: two episodes of mild conditions (the Bølling (13,000–12,000 yr BP) and Allerød (11,800–11,000 yr BP) interstadials) separated by two cold periods (the Older Dryas (12,000–11,800 yr BP) and the Younger Dryas (11,000–10,000 yr BP) stadials). In Britain, the Late-glacial is divided into the *Windermere*

Interstadial (13,000–11,000 yr BP) and the *Loch Lomond Stadial* (11,000–10,000 yr BP) (Lowe and Gray, 1980). From the European mainland, a sequence of two interstadials (*Bølling* and *Allerød*) was separated by the *Older Dryas* short cold episode, and the Allerød was followed by the *Younger Dryas Stadial* (Fig. 5.9).

The stratigraphic sequence of Late-glacial *chronozones* has, however, been hampered by problems with accuracy of radiocarbon dating during this time interval. In northwest Europe, the chronozones are radiocarbon-dated *biozones*, but biozones are commonly time-transgressive as a response to climatic change, which are geographically and temporally diachronous. The Younger Dryas chronozone spans the time interval between 11,000 and 10,000 yr BP, but the Younger Dryas biozone is part of a stratigraphic sequence characterized by cold fossil assemblages, and these two rarely coincide.

Based on De Geer's (1912) varve investigations, the Swedish Time Scale has been divided into 'postglacial', 'finiglacial', and 'gotiglacial' parts. About 9266 'postglacial' varves were deposited during the Holocene along the river Ångermanälven in central Sweden. The 'finiglacial' (early Holocene) varves have been connected to the oldest 'postglacial' varves, and this series consists of 1191 varves (Strömberg, 1989). These varves have then been connected to the youngest Late Weichselian ('gotiglacial') varves through overlapping varve diagrams (e.g. Brunnberg, 1995). To test whether the visual correlations were statistically correct, Holmquist and Wohlfarth (1998) used cross-correlation measures with overlapping varve diagrams from two local varve chronologies established in southeastern Sweden. Of a total of 363 analysed connections, only 78 fulfilled the statistical requirements for a perfect match. In 96 cases, the statistical approach suggested alternative links. In addition, they found that 179 correlations were not statistically valid and that 11 overlaps were too short to allow valid cross-correlation. The authors therefore suggested statistical analysis of the varve chronology links before the Swedish varve chronology can be regarded as a valid, high-precision time-scale.

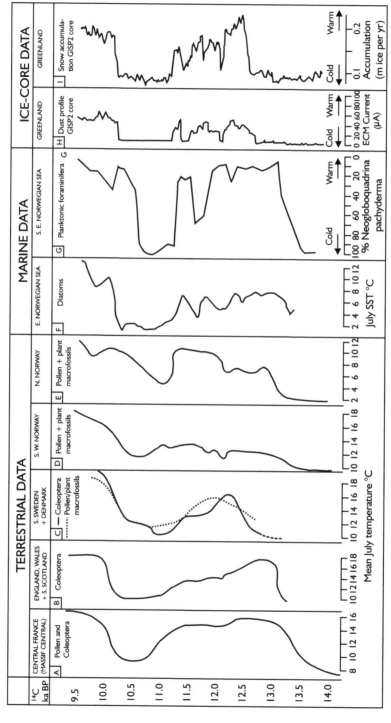

FIGURE 5.9 Climate development in Europe and adjacent areas of the North Atlantic based on terrestrial, marine and ice-core data. (Modified from Walker, 1995)

New and advanced dating possibilities, including accelerator mass spectrometry (AMS) measurements on terrestrial macrofossils, have demonstrated the problems inherent in the present subdivision of the Late-glacial period (e.g. Ammann and Lotter, 1989). Atmospheric ^{14}C variations during the Late-glacial and early Holocene periods led to significant differences in radiocarbon and calendar years, and plateaux of constant radiocarbon age at ca. 12,700, 10,400 and 10,000 yr BP (Ammann and Lotter, 1989). The present model of radiocarbon-dated calendar-year time-scales is based on dendrochronology, varve chronology and U/Th dates on corals (Wohlfarth, 1996). It has therefore been suggested that the subdivision into chronozones should be abandoned and that isotope signals should be used instead (Broecker, 1992). This was suggested because plateaux of nearly constant radiocarbon age coincide with periods of rapid climate change (e.g. Becker et al., 1991), making it impossible to date the transitions precisely using the radiocarbon method. In addition, AMS radiocarbon dates obtained from terrestrial plant macrofossils are several hundred years younger than the corresponding bulk samples from the same stratigraphic levels, mainly due to hardwater effects (e.g. Wohlfarth et al., 1993). The terminological and chronological complexity arising from use of the radiocarbon method leads to further complications.

A series of 70 AMS radiocarbon dates has been used to date the Younger Dryas/Holocene (YD/H) transition in the lacustrine sediments of Kråkenes Lake, western Norway to 11,530 +40/-60 cal. yr BP (Gulliksen et al., 1998). This age estimate for the YD/H transition is in close agreement with evidence presented from the Greenland GISP2 and GRIP ice cores, German pine series, European lake sediments and Baltic varves, indicating that the YD/H transition in the North Atlantic region occurred between 11,600 and 11,500 cal. yr BP.

During the last two decades, the Late-glacial climatic and glacial histories of the North Atlantic region have been interpreted in term of shifts in the North Atlantic oceanic Polar Front (Ruddiman and McIntyre, 1981; Karpuz et al., 1993). Variations in the position of the Polar Front have been assumed to reflect oceanographic changes, such as the thermohaline ocean conveyor belt (Broecker and Denton, 1990), a sudden influx of freshwater from melting ice sheets (Broecker et al., 1989), and energy transfer in connection with the formation of deepwater in the North Atlantic (Broecker et al., 1985). Several feedback mechanisms operate, like the effects of the Late Cenozoic ice sheets and sea-ice cover on atmospheric and surface ocean circulation (Kutzbach and Wright, 1985; Karpuz and Jansen, 1992). The Greenland ice cores show climatic changes roughly in phase with the fluctuations of the oceanic Polar Front (e.g. Dansgaard et al., 1989). Both the ice-core and marine records show that large-scale climatic changes occurred very rapidly in the North Atlantic region.

5.5 Variations of local glaciers during the last glaciation

For the Colombian Andes, a radiocarbon-dated framework of pre-Late-glacial maximum was presented by Helmens et al. (1997). During each of three pre-LGM advances, glaciers extended beyond the LGM. In other parts of the Andes, in New Zealand, and in some Eurasian mountain ranges, glaciers advanced farther earlier in the last glacial cycle than at the LGM (Gillespie and Molnar, 1995; Bondarev et al., 1997; Clapperton et al., 1997; Fitzsimons, 1997; Clapperton, 1998). The interval of greatest glacier extent is not dated, but it is commonly assumed to correspond to the cooling recorded in isotope stage 4, with an approximate duration of 10,000 years. As a comparison, isotope stage 2 lasted for about 16,000 years, during which the northern hemisphere ice sheets grew to their greatest extent. Assuming that sea-surface temperatures were not as low during isotope stage 4 as during isotope stage 2, but that precipitation was generally higher, it is conceivable that local glaciers grew to their maximum extent during isotope stage 4 (Clapperton, 1997).

The timing and extent of local glaciers during the LGM are uncertain and variable, and the concept of a single LGM is misleading

(Clapperton, 1997). For the Colombian Andes, Helmens *et al.* (1997) reported advances at 23,900–19,500 yr BP, and between 18,000 and 15,500 yr BP. Glaciers in the more arid Andes and Bolivia, on the other hand, reached their greatest extent after ca. 13,300 yr BP (Clapperton *et al.*, 1997). In southern Chile, Lowell *et al.* (1995) dated glacier advances at 23,000, 21,000 and 17,000 yr BP. In New Zealand, the LGM glaciers reached their maximum at 23,300–18,000 yr BP (Fitzsimons, 1997). In the European Alps, glacier expansion to the Würm (LGM) maximum occurred after 24,000 yr BP (van Husen, 1997).

In some regions, local glaciers underwent glacier readvance after the LGM and prior to the 13,500 yr BP warming. In the European Alps, a glacier advance (Gschnitz) culminated at ca. 16,000–14,000 yr BP (van Husen, 1997). In Scotland, the Wester Ross Advance occurred between 17,000 and 12,800 yr BP (Benn, 1996; Ballantyne, 1997) and may have been contemporaneous with Heinrich layer 1 in the North Atlantic at 15,000–14,000 yr BP (Bond and Lotti, 1995). In the Southern Alps of New Zealand, glaciers advanced at ca. 15,000–14,000 yr BP (Fitzsimons, 1997) and this advance was probably simultaneous with a glacier advance in southern Chile (Clapperton, 1995), with a maximum extent at 14,900–13,900 yr BP (Lowell *et al.*, 1995). Glaciers in SE Peru experienced their maximum isotope stage 2 positions at 14,000–13,900 yr BP (Mercer and Palacios, 1977; Clapperton, 1993). In the interval 18,000–13,500 yr BP, many glaciers seem to have advanced between 14,500–13,500 yr BP, and this readvance was terminated by abrupt global warming at 13,500 yr BP.

The interval between 13,500 and 10,000 yr BP was characterized by rapidly receding glaciers as a response to global warming, and most of the main valleys in Scotland, the Alps, southern Chile and New Zealand were deglaciated by ca. 13,000 yr BP (Sutherland, 1984; Porter, 1981b; Lowell *et al.*, 1995). In Columbia and Bolivia, however, Helmens *et al.* (1997), Clapperton (1995) and Clapperton *et al.* (1997) reported the expansion of local glaciers between 13,500 and 12,500 yr BP. A cooling at around 12,500 yr BP is reflected in North

Atlantic pollen diagrams (Levesque *et al.*, 1993; Lowe *et al.*, 1995), in glacier expansion in the Alps (the Daun Stadial) and Scandinavia (the Older Dryas stadial), and in Greenland ice cores (Alley *et al.*, 1993). Between 11,000 and 10,000 yr BP, the North Atlantic region in particular was plunged into a severe cooling (the Younger Dryas). Throughout Europe local glaciers readvanced, and outlet glaciers along the western margin of the Scandinavian ice sheet expanded by tens of kilometres (Mangerud, 1991). In recent literature, the question of whether the Younger Dryas event was global or not has been a continuous issue. Compared with the evidence presented from Europe, however, the amplitude of glacier expansion outside the amphi-Atlantic region was small. However, in the Colorado Front Range, Menounos and Reasoner (1997) found evidence in lake sediments for a Younger Dryas glacier episode.

5.6 Holocene glacier and climate variations

5.6.1 Canada

The Little Ice Age was the most extensive Neoglacial glacier advance in the Canadian Rocky Mountains (Luckman *et al.*, 1993). Evidence of earlier, less extensive Neoglacial advances is based on wood recovered from several glacier forefields. Three radiocarbon dates, ranging between 8230 and 7550 yr BP, obtained from wood flushed out of Athabasca Glacier, and two dates from Dome Glacier ranging between 6380 and 6120 yr BP, indicate that forests occurred upvalley of present glacier fronts during the Hypsithermal. Radiocarbon dates from detrital and *in situ* logs indicate that forests were overridden by glaciers between 3100 and 2500 yr BP. This advance, termed the Peyto Advance, did not extend beyond the Little Ice Age maximum position. The earliest Little Ice Age advance is dated to ca. 900–600 yr BP.

Prior to the deposition of the Mazama tephra 6800 yr BP, a minor glacier readvance (the Crowfoot Advance) left deposits in the Rockies and the interior of British Columbia

(Osborn and Luckman, 1988). The first Neo-glacial advances took place 6000–5000 yr BP. Other advances occurred between 4000–3000 yr BP and at about 2500 and 1800 yr BP. The Little Ice Age expansion, which started shortly after 900 yr BP, culminated in the eighteenth and nineteenth centuries.

5.6.2 USA

In the American Cordillera, a Late-glacial or early Holocene glacier readvance or stillstand deposited moraines about 1–3 km beyond present glacier fronts (Thompson Davis, 1988). The earliest dated Neoglacial advances occurred at about 5000 yr BP. In most mountain ranges of the western USA, unweathered, sharp-crested moraines adjacent to modern ice margins, or near cirque headwalls without glaciers at present, date to the Little Ice Age of the last several centuries.

5.6.3 The Arctic

A review of the Holocene glaciation record in Alaska (Calkin, 1988) suggested that glacier fluctuations between Arctic, central interior and southern maritime Alaska were mostly synchronous. There is evidence of glacier expansion between 7600 and 5800 yr BP, and between 5800 and 5700 yr BP. A significant increase in glacier activity began at 4400 yr BP. During the Holocene glacial maxima, glaciers covered Glacier and Lituya Bays and the fjord of Prince William Sound. A minor glacier recession has been recorded at about 2000 yr BP. The widespread Little Ice Age glacier advances were initiated at approximately 700 yr BP. In the Arctic, glacier advances of AD 1600 dominated those of the late 1800s.

In a study by Evans and England (1992), they reported that indicators of sea-ice conditions on northern Ellesmere Island suggest that the early Holocene was a period of considerable open water. Geomorphological evidence shows that the ice shelves are presently breaking up and melting in response to recent warming. Several large glaciers in that region are still advancing in response to the mid-Holocene climatic deterioration. Some

glaciers display evidence of dual advances which may reflect mid-Holocene and Little Ice Age accumulation. Other evidence of Little Ice Age cooling and recent warming includes perennial snowbank retreat and fluvially eroded ice-wedge polygons near sea-level.

Bradley (1990) reviewed the Holocene record of climatic change on the Queen Elizabeth Islands. He found that temperatures were highest during the early to mid-Holocene. Temperatures declined from approximately 3000 yr BP, culminating in low temperatures from 100–400 yr BP. This coldest period during the Holocene resulted in glacial advances to post-glacial maximum positions. Since 1925 AD there has been a pronounced temperature increase which has led to negative mass balance on glaciers and ice sheets.

Porter (1989) presented a 2000-year chronology of fluctuations of the tidewater glacier in Icy Bay. The outermost moraine complex at the mouth of the bay dates from AD 400 to 850. Subsequent glacier retreat led to deglaciation of the fjord by AD 1000. At the onset of the Little Ice Age (thirteenth century) the glacier expanded and reached its maximum in the early nineteenth century. A study by Wiles and Calkin (1994) from Kenai Mountains, Alaska, revealed three major intervals of Holocene glacier expansions: 3600 yr BP, AD 600, and during the Little Ice Age.

5.6.4 Greenland

Investigations along the western Greenland ice sheet margin in the region of the Jacobshavn Isbræ indicate that the glacier was at least 15 km behind the present position between 4700–2700 cal. yr BP (Weidick *et al.*, 1990). Gordon (1980) demonstrated that cirque and valley glaciers in western Greenland have advanced since 1968 simultaneously with the retreat of larger valley and icefield outlet glaciers.

5.6.5 Iceland

Historical records indicate that the maximum Holocene extent of Fjallsjökull and Breidamer-kurjökull was in the latter half of the nineteenth century (Kugelmann, 1991). The sequence of Holocene development in Iceland can be summarized as: (1) Late Weichselian and early Holocene glaciation; (2) Holocene non-glacial conditions and prehuman colonization; (3) Holocene non-glacial conditions with human occupation; (4) Little Ice Age glaciation (Norddahl, 1990; Ingólfsson, 1991). There are, however, uncertainties concerning the timing of the wastage of the Weichselian glaciers, and the timing and extent of glacial readvances between the Late Weichselian deglaciation and the Little Ice Age glaciation (Gudmundsson, 1997). However, stratigraphic evidence from Skalafellsjökull, SE Iceland, and Eyjabakkajökull, eastern Iceland, suggests no glacier advance during this interval (Sharp and Dugmore, 1985).

A study of the Holocene record of glacier fluctuations from northern Iceland (Stötter, 1991) indicated two glacial advances during the time interval 6000–4800 yr BP. Dugmore (1989) used tephra layers interbedded with soils to date Holocene glacier fluctuations in southern Iceland. This study demonstrated that a large ice mass existed in the mid-Holocene in Iceland, because Sólheimajökull extended up to 5 km beyond its present limits between 7000 and 4500 yr BP. Major advances also culminated before 3100 yr BP and between 1400–1200 yr BP. In the tenth century the glacier was also longer than during the Little Ice Age (AD 1600–1900). In contrast, some other glaciers reached their maximum Holocene extent during the Little Ice Age. The anomalous behaviour may be the result of changes in the catchment areas over the last 5000 years. Recently, Stötter et al. (1999) suggested that Holocene glacier advances in northern Iceland took place at around 4700, 4200, 3200–3000, 2000, 1500 and 1000 yr BP.

A study at Fjallsjökull, SE Iceland, indicated a mid-Holocene glaciation ending at about 4500 yr BP (Rose et al., 1997). The Little Ice Age glaciation occurred ca. AD 1850–1965. At Fjallsjökull this glaciation caused glaciotectonic deformation and subsequent erosion of the mid-Holocene land surface.

5.6.6 New Zealand

In the central Southern Alps of New Zealand, glacier retreat exposed sections in lateral moraines which made possible studies of moraine stratigraphy and genesis (Gellatly et al., 1988). Radiocarbon dating of buried wood and soils indicates that glaciers in New Zealand expanded at about 5000, 4500–4200, 3700, 3500–3000, 2700–2200, 1800–1700, 1500, 1100, 900, 700–600 and 400–100 yr BP.

5.6.7 South America and Antarctica

Different dating approaches indicate that glaciers readvanced significantly only during the last 5000 yr BP (Clapperton and Sugden, 1988). In South America there is evidence for four main periods of Neoglacial advance: 5000–4000, 3000–2000, 1300–1000 yr BP, and fifteenth to late nineteenth centuries. Minor glacier advances may have occurred at ca. 8400, 7500 and 6300 yr BP.

Different Antarctic environments produce interesting contrasts. Some glaciers in McMurdo Sound (East Antarctica) are more extensive now than during the global glacial maximum at 18,000 yr BP. Radiocarbon evidence from the South Shetland Islands indicates two main Holocene glacier advances, the most extensive peaking in the twelfth century.

At South Georgia, Clapperton et al. (1989) used glacial geomorphology, slope stratigraphy, and analyses of environmental indicators in peat and lacustrine cores to present a Holocene record. At low altitudes plant growth had begun by 9700 yr BP. Fairly cool conditions prevailed until 6400 yr BP, followed by a period from 5600 to 4800 yr BP when conditions were slightly warmer than at present. Periods of climatic deterioration occurred at around 4800–3800 yr BP, 3400–1800 yr BP, and during the last 1400 years. The most extensive Holocene glacier advance on South Georgia culminated slightly before 2200 yr BP.

5.6.8 Mount Kenya and other East African mountains

Many of the East African mountains are covered with late Pleistocene and late Holocene glacial sediments which are poorly dated. Only on Mount Kenya is there a chronology emerging based on dating. During the Pleistocene, glacier fluctuations appear to have been broadly synchronous in the East African mountains. Terminal moraines on Mount Kenya extend to 3200 m in most drainage basins and yield minimum radiocarbon dates of approximately 15,000 yr BP (Mahaney, 1988). Recessional moraines at 3700–4200 m appear to date from the Late-glacial, while younger Neoglacial moraines are either Little Ice Age or older late Holocene deposits. The Neoglacial sequence on Mount Kenya is younger than 1000 yr BP, while on other East African mountains glaciers may have existed during the last 2000 yr BP.

Lacustrine sedimentary evidence indicates that glaciers occupied the southwestern cirques on Mount Kenya during much of the last 6000 years (Karlén, 1998; Karlén *et al.*, 1999). Pro-glacial lacrustrine sediments obtained from Hausberg Tarn suggest six major periods of glacier advances at about 5700, 4500–3900, 3500–3300, 3200–2300, 1300–1200, and 600–400 cal. yr BP. The 5700 cal. yr BP advance reached approximately 1 km beyond the Little Ice Age moraines.

5.6.9 The Alps

Röthlisberger (1986) summarized the Holocene glacier fluctuations in the Alps. According to his data compilation, the glaciers were in advanced positions at 8600, 8200, 7500, 6400, 6100, 5200, 4800, 4500, 3600–3000, 2800, 2100, 1100, 600, 400 and 180 yr BP. On the Swiss Plateau and at timberline in the Alps, Haas *et al.* (1998) found, on the basis of palaeobiological evidence, eight synchronous pre-Roman cold phases at 9600–9200, 8600–8150, 7550–6900, 6600–6200, 5350–4900, 4600–4400, 3500–3200 and 2600–2350 yr BP.

Cores recovered from the proglacial Lake Silvaplana in the Swiss Alps (Leemann and Niessen, 1994) showed that glacial varves were deposited during glacial retreat in the early Holocene until 9400 yr BP. Glacial activity was absent or negligible within the catchment between 9400 and 3300 yr BP. Maximum varve

thickness, interpreted as reflecting the size of the glaciers in the catchment, was observed between AD 1790 to 1870.

Grove (1997) summarized the history of Holocene glacier variations in the Alps. The sequence of deglaciation in the Austrian Ötztal is documented by peats dated at 10,000 BP within the Egesen moraines, correlated with the Younger Dryas in NW Europe. A glacier advance took place in the Swiss and Austrian Alps at around 9300 yr BP, termed the Schlaten advance period between 9400 and 9000 yr BP. The period between 9000 and 6000 yr BP was characterized by episodes of recession interrupted by glacier advances between 6600 and 6000 yr BP, between 7700 and 7300 yr BP, and between 8400 and 8100 yr BP. The interval between 6000 and 4600 yr BP was generally warm, but broken by glacier advances at around 4600 and 4200 yr BP. A warm climate prevailed between 4600 and 3600 yr BP, but during the period 3600–3000 yr BP a significant glacier advance occurred (the Löbben Advance). The period from 3000 to 1100 yr BP was warm and characterized by glacier contraction. Between 900 and 800 yr BP glaciers readvanced. During the Medieval period (ca. AD 900–1300) glaciers retreated and reached positions comparable to those in the late twentieth century.

In the Italian Alps, glaciers were reduced to their present size, or even smaller, after deglaciation at about 9000 yr BP (Orombelli and Mason, 1997). In the western Italian Alps, Rutor Glacier had its terminus upvalley of the present position during most of the period between 9000 and 5000 yr BP. Subsequent to 5000 yr BP, several Neoglacial advances occurred, the most pronounced between 3000 and 2500 yr BP. The first Little Ice Age glacier advance was recorded as post-Medieval floods from the Rutor Glacier. The maximum Little Ice Age extent was reached during the seventeenth and eighteenth centuries. The majority of the glaciers reached their maximum Holocene extent around AD 1820 or 1850. Minor readvances took place at approximately AD 1890, 1920–1925 and 1970–1980. The glaciated area has been reduced by around 40 per cent since the last century.

Dendrochronological analyses of fossil larches in the forelands of Grosser Aletch

Glacier show that the glacier advanced during the Löbben cold period at around 3200–3100 yr BP, during the Göschenen cold period I at approximately 2800–2700 and 2400–2300 yr BP (Holzhauser, 1997). Grosser Aletch Glacier was 600–1000 m less advanced than at present between 3200 and 2400–2300 yr BP. Dendroclimatological research proved that the Gorner Glacier advanced during the fourteenth century and that the Medieval warm period lasted from the eighth to the end of the thirteenth century (Fig. 5.10).

Tree trunks and wood fragments in minerotrophic peat accumulated in the outwash plain of Unteraargletscher, Switzerland, have been radiocarbon-dated to represent Holocene retreat phases of the glacier (Hormes et al., 1998). The radiocarbon dates suggest that the glacier was at least several hundred metres less advanced at around 8100–7670, 6175–5780, 4580–4300 and 3380–3200 yr BP. The warmest and driest period occurred between 4100–3600 yr BP. Growth of peat between 3800 and 3600 yr BP, was, however, attributed to more humid conditions. Based on the chronology, the authors suggested an approximately 2000-year cyclicity of tree and/or peat growth in the study area.

5.6.10 China

In China, the last 10,000 years have been climatically divided into three parts. Humid conditions prevailed in the early Holocene (10,000–7500 yr BP), followed by the climatic optimum at 7500–3000 yr BP. At about 3000 yr BP, temperatures dropped, causing numerous glacier advances over the Tibetan Plateau.

5.6.11 Scandinavia

Karlén (1988) presented evidence of significant Neoglacial episodes in northern Scandinavia at about 7500, 5100–4500, 3200–2800, 2200–1900 and 1500–1100 yr BP, while less extensive glacier readvances, for which there is less evidence, occurred at 6300, 5600, 2500, 940, 600–560 and 380 yr BP. Pre-Little Ice Age moraines have been mapped around Jostedalsbreen in western Norway, the largest ice cap on mainland Europe. The moraines have been radiocarbon-

dated to 9100 yr BP, and termed the Erdal event by Nesje et al. (1991), Nesje and Kvamme (1991) and Nesje (1992). Similar moraines have also been reported from the Hardangerjøkulen and Jotunheimen areas. At Hardangerjøkulen, a short-lived glacier phase is dated at 7600 yr BP (8200 cal. yr BP) (Nesje and Dahl, 1991a; Dahl and Nesje, 1994, 1996) and termed the Finse event. Radiocarbon dates indicate that glaciers on the northern part of the Hardangerjøkulen Plateau were totally melted between 7560 and 6285 yr BP. In northern Scandinavia, at Jostedalsbreen, and at the northern part of Hardangerjøkulen, glacier readvances occurred between 6300 and 5300 yr BP (Karlén, 1988; Nesje et al., 1991; Dahl and Nesje, 1994, 1996). Glaciers at the northern part of Hardangerjøkulen were small or totally absent between 5280 and 4830 yr BP. At the northern sector of Hardangerjøkulen, the period from 4830 to 3790 yr BP was characterized by minor, high-frequency glacier oscillations. Hardangerjøkulen has existed continuously from 3790 yr BP to the present (Dahl and Nesje, 1994, 1996). This is also in close accordance with a date of 3710 yr BP for evidence of significant Neoglacial expansion in the Jostedalsbre region (Nesje et al., 1991). At the southwestern margin of the Jostedalsbre ice cap, Nesje and Dahl (1991b) showed that a cirque glacier isolated from the main ice cap was present between 2595 and 2360 yr BP, 2250 and 2150 yr BP, 1740 and 1730 yr BP, and subsequent to 1430 yr BP.

Two radiocarbon-dated soils on the surface of Blåisen and Midtdalsbreen, at the northern margin of Hardangerjøkulen, demonstrate that both outlet glaciers were behind the modern terminal position until about 1500–2000 yr BP (Dahl and Nesje, 1994). Two minor glacier oscillations also occurred at about 1100 and 800 yr BP (Nesje and Dahl, 1991a; Dahl and Nesje, 1994, 1996), while the Little Ice Age is dated to have occurred since 575 yr BP (cal. yr AD 1300–1420).

Karlén and Matthews (1992) and Matthews and Karlén (1992) found evidence of glacier expansions in the Møre/Jostedalsbreen/Jotunheimen regions during the following intervals: 6400–5900, 3400–3000, 3000–2600, 2500–2300, 2200–2100, 1600–1400 and after 1000 yr BP.

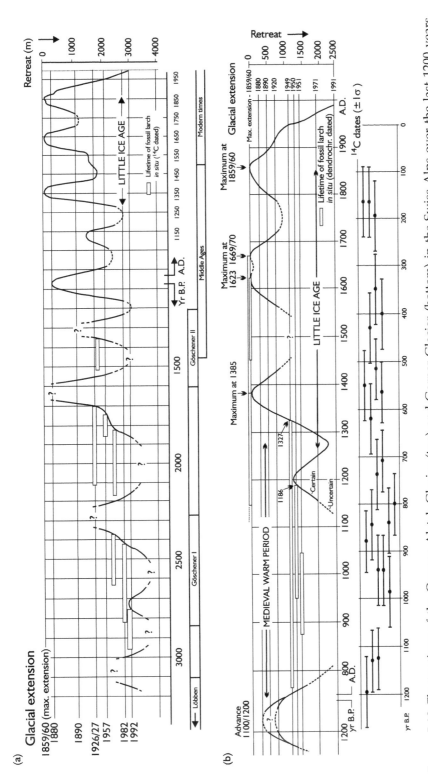

Figure 5.10 Fluctuations of the Grosser Aletch Glacier (top) and Gorner Glacier (bottom) in the Swiss Alps over the last 1200 years. (Adapted from Holzhauser, 1997)

Matthews (1991) presented convincing evidence from radiocarbon-dated buried soils for a late Neoglacial maximum within the last 400–600 years in the Jostedalsbreen/Jotunheimen region. Griffey and Worsley (1978) found no evidence for any extensive glacier episode after the deglaciation of the continental ice sheet between approximately 9000 and 3000 yr BP. Two significant pre-Little Ice Age periods of glacier advance took place in the Okstindan region (Griffey and Worsley, 1978). The first was dated to between 3000–2500 yr BP, and the second to between the eighteenth century and 1600 yr BP, with a tentative age of 1250–1000 yr BP.

In front of glaciers on Lyngshalvøya, northern Norway, up to four Neoglacial moraines occur (Ballantyne, 1990), representing five episodes of glacier expansion, one of which predated the Little Ice Age. Lichenometric, dendrochronological and historical evidence indicates that the oldest Little Ice Age moraines date to the mid-eighteenth century, and the youngest to AD 1910–30. At nine small glaciers, the AD 1910–30 moraine represents the Neoglacial maximum.

5.6.12 Tatra Mountains

In the Tatra Mountains, about ten small glaciers (glacierets) occur (Jania, 1997). The absence of typical glaciers makes reconstruction of Holocene glacier fluctuations difficult. Geomorphological evidence suggests a total melt of glaciers in the highest cirques subsequent to 8300 yr BP. A fossil soil found beneath one of the glacierets indicates that the snow patches did not exist in the Tatras during the Atlantic period. During the Little Ice Age, however, glacierets and snow patches were considerably larger than during recent decades. During the last 20 years, the number of snow patches has decreased and those remaining are also reduced in size and thickness (Jania, 1997).

5.6.13 Jan Mayen

On the island of Jan Mayen ($373\,km^2$), there have been at least two periods of glacier advance in the Holocene (Anda *et al.*, 1985).

The first may have occurred at approximately 2500 yr BP, while some glaciers reached their maximum extent around AD 1850. Subsequently the glaciers experienced an oscillating retreat, however, with a significant expansion around AD 1960.

5.7 Neoglacial glacier variations

The term 'neoglaciation' was introduced by Porter and Denton (1967). The term refers to the readvance or formation of glaciers after their minimum extent during the early Holocene. The start of neoglaciations varies significantly from region to region (e.g. Denton and Karlén, 1973; Grove, 1988). On Baffin Island, Arctic Canada, the glacier margins were behind present positions at 5000 yr BP, with short readvances before about 3200 yr BP. Neoglacial advances in North America are reported from the Canadian Cordillera, the US Rocky Mountains, the Brooks Range in Alaska, and the Torngat Mountains in Labrador. The Neoglacial chronologies have been reconstructed from marginal moraine sequences and material subsequently overrun by glaciers. Investigations of the present glaciers in Scandinavia suggest that most, if not all, glaciers were totally melted once or several times during the Holocene (e.g. Karlén, 1988). In New Zealand, stratigraphic sequences provide a pattern of Neoglacial variations in close agreement with the European records (Gellatly *et al.*, 1988). The Neoglacial history of the Antarctic ice sheet is characterized by relatively small fluctuations, although much research is still to be done.

5.8 Little Ice Age glacier variations

During the last few centuries glaciers advanced on all continents, indicating that the Little Ice Age was a global phenomenon. In the European Alps, the main advance around AD 1850 was preceded by another of similar magnitude around AD 1300. Thirteenth to fourteenth-century advances in Scandinavia and North America are not well documented;

however, the evidence for such advances in the Himalayas and New Zealand are better documented (Grove, 1988). In the European Alps, the initiation of the main advances led to glacier expansion close to the Little Ice Age maxima around 1600. In southern Norway, on the other hand, the glacier expansion started around the mid-seventeenth century. In Iceland, sea ice expanded around the middle of the century, although the glaciers started to advance at the end of the seventeenth century. Between AD 1600 and 1850 the glaciers in the Alps repeatedly reached almost the same terminal positions. In Norway, the moraines formed since the mid-eighteenth century are separated from each other, with successively younger moraines toward the present glacier snouts. Closely spaced moraines formed by advances of the New Zealand glaciers on the eastern side of the Alps date to AD 1650 and 1885. Since the late nineteenth and the twentieth centuries, most glaciers in the world have been retreating, interrupted only by minor readvances. In the 1990s, however, maritime glaciers in western Scandinavia have advanced significantly as a response to increased winter precipitation (Nesje et al., 1995).

5.8.1 Iceland

Vatnajökull (8538 km^2) is the largest ice cap in Europe and the glacier rests on a series of active volcanoes centred around Grimsvötn. Several of the western and northern outlet glaciers are surging glaciers. Historical evidence suggests that Vatnajökull was quite large by the end of the seventeenth century, causing damage to farms and pastures. Between AD 1690 and 1710 the Vatnajökull outlet glaciers advanced rapidly. In the subsequent decades the glacier termini were stationary or fluctuated a little. Around 1750–1760 a significant readvance occurred, and most of the glaciers are considered to have reached their maximum Little Ice Age extent at that time (e.g. Grove, 1988). During the mid-eighteenth to late nineteenth centuries, glaciers at the southern side of Vatnajökull were quite extensive. During the twentieth century, however, glaciers

retreated rapidly. As an example, Breidamerkurjökull, which comprises about 14 per cent of Vatnajökull, decreased in volume by about 49 km^3 between 1894 and 1968, while the whole glacier diminished in volume by between 268 and 350 km^3 (8–10 per cent) (Grove, 1988). Length variations of south- and south-east flowing outlet glaciers from Vatnajökull between 1930 and 1995 are shown in Fig. 5.11.

During the Little Ice Age, Myrdalsjökull (700 km^2) and Eyjafjallsjökull formed one ice cap, which in the middle of the twentieth century separated into two ice caps (Grove, 1988). Myrdalsjökull covers Katla, the second most active volcano in Iceland. Therefore, volcanic eruptions from Katla are accompanied by floods (jökulhlaups). The coastal settlements south and east of Myrdalsjökull have suffered from volcanic eruptions, floods, glacier advances and avalanches. Length variations of Myrdalsjökull and Eyjafjallsjökull between 1930 and 1995 are shown in Fig. 5.12, p. 146.

The most detailed information about glacier variations exist from Sólheimajökull, a long outlet glacier in the southwest. A Danish map from 1904 shows that the Sólheimajökull terminus had retreated to an altitude of about 100 m a.s.l. The eastern glacier front retreated by about 200 m between 1883 and 1904, whereas the western terminus was stationary. Between 1930 and 1937 the glacier thinned and retreated with a mean rate of 30–40 m yr^{-1}. Since 1930, the glacier fronts of several outlet glaciers have been measured annually. The glaciers retreated until the first part of the 1960s, after which the glaciers have started to advance.

Drangajökull (166 km^2) is a small ice cap in northwest Iceland. By the end of the seventeenth century, Drangajökull advanced across farmland, and during the mid-eighteenth century the outlet glaciers were the most extensive known since the surrounding valleys were settled. The available historical evidence suggests that before a significant advance that occurred around 1840, there seems to have been a small retreat. After the mid-nineteenth century advance, glaciers retreated significantly. Length variations of Drangajökull and

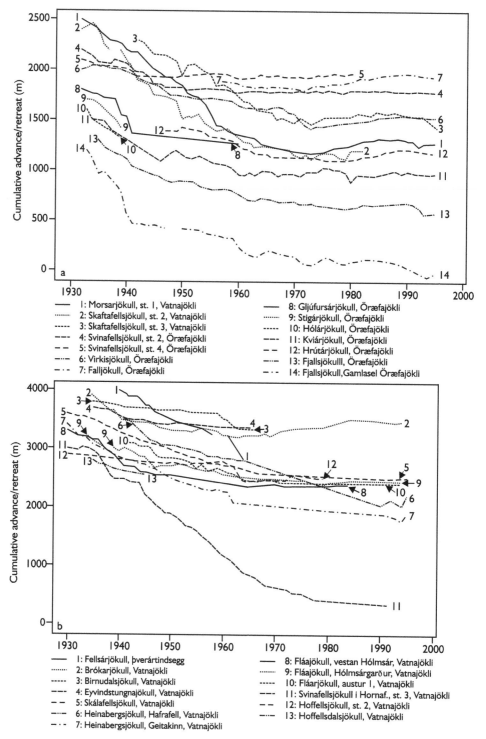

FIGURE 5.11 Length variations of south- (top) and southeast-flowing (bottom) outlet glaciers from Vatnajökull between 1930 and 1995. (Modified from Sigurdsson, 1998)

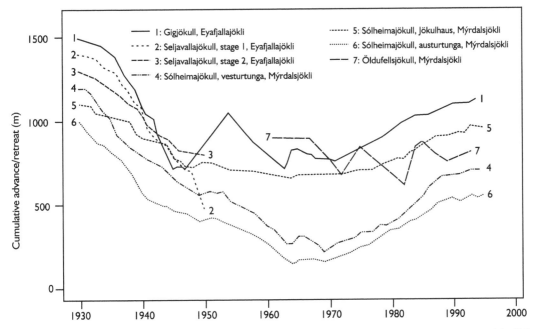

FIGURE 5.12 Length variations of Myrdalsjökull and Eyjafjallsjökull between 1930 and 1995. (Modified from Sigurdsson, 1998)

Snæfellsjökull between 1930 and 1995 are shown in Fig. 5.13.

LANDSAT images obtained between 1973 and 1992, combined with field observations, were used to measure changes in terminal positions of outlet glaciers from Vatnajökull, Iceland (Williams *et al.*, 1997). The largest changes during the 19-year period occurred in the large, lobate, surge-type outlet glaciers along the southwestern, western and northern margins of Vatnajökull, experiencing a glacier retreat of up to about 2 km during the study interval.

5.8.2 Scandinavia

5.8.2.1 Jostedalsbreen

Farms in inner Nordfjord were severely damaged by glacier advances of the western outlet glaciers from Jostedalsbreen (and associated avalanches, rockfalls and landslides) in the seventeenth and eighteenth centuries (Grove and Battagel, 1983; Grove, 1988). Evans *et al.* (1994) claimed from studies

in the Sandane area in Nordfjord that moraines were formed in the thirteenth to fourteenth centuries. Matthews *et al.* (1996) made a field survey in the study area of Evans *et al.* (1994), and concluded that the evidence of pre- or early Little Ice Age moraines cannot be supported. Historical documents and lichenometry demonstrate that several of the outlet valley glaciers from Jostedalsbreen reached their maximum position during the mid-eighteenth century (Grove, 1988; Bickerton and Matthews, 1993). Dahl and Nesje (1992) calculated from a reconstructed Little Ice Age cirque glacier in inner Nordfjord that winter precipitation was reduced to about 90 per cent of present values, with a corresponding mean ablation-season temperature depression of approximately 1.5°C compared with the present. The most representative Little Ice Age ELA depression in the Jostedalsbre region is calculated to be about 150 m (Nesje *et al.*, 1991), while Torsnes *et al.* (1993) calculated the average Little Ice Age ELA depression for 20 outlet glaciers from the Jostedalsbreen ice cap as 80 m, by means of the AAR approach.

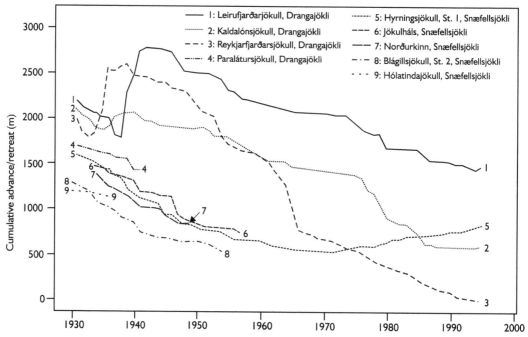

FIGURE 5.13 Length variations of Drangajökull and Snæfellsjökull between 1930 and 1995. (Modified from Sigurdsson, 1998)

In the valleys surrounding Jostedalsbreen, the farming economy was, and still is, based on pastoralism with summer pastures (sæter) in adjacent valleys and on the valley sides. From a letter of AD 1340, a disaster of some kind seems to have occurred in the Jostedalsbreen region in the first part of the fourteenth century (Grove, 1972, 1988). In this letter, farms which were affected were listed and tax reductions were ordered. The abandonment of farmland around Jostedalsbreen took place before the Black Death, which caused a dramatic decline in population. In the 1970s an area of fresh till overlying humus and plant remains was exposed at Omnsbreen, north of Finse, central southern Norway (Elven, 1978). Radiocarbon dating of the plant material suggested a glacier advance in the fourteenth century. Historical evidence, and radiocarbon dating of palaeosols buried by the outer Neoglacial moraines suggest that most glaciers in southern Norway expanded significantly from the seventeenth century onwards. The first reliable evidence of damage to farmland caused by advancing glaciers in

Scandinavia comes from Krundalen, a western tributary to Jostedalen east of Jostedalsbreen. In a brief account dated 1684, two farmers pleaded that they were not able to pay their taxes because their pastures had been covered by an advancing glacier. Historical evidence suggests that the glaciers in Jostedalen advanced rapidly during the late 1600s and early 1700s. Between AD 1710 and 1735, Nigardsbreen, an eastern outlet glacier of Jostedalsbreen, advanced 2800 m: an average advance rate of 112 m year^{-1}. As the valley outlet glaciers expanded, the damage caused to farmland and pastures led to local investigations (*'avtaksforretninger'*) and courts of inquiry. These *avtaksforretninger* led to the accumulation of many documents, which today give insight into the suffering of the people who lived in the vicinity of the glacier. Mattias Foss, the vicar in Jostedalen at that time, wrote in 1743 (translation in Grove, 1988): 'the glacier had carried away buildings; pushing them over and tumbling them in front of it with a great mass of soil, grit and great rocks from the bed and had crushed the buildings to very small pieces which are still to be seen, and the man who lived there has had to leave his farm in haste with his people and possessions and seek shelter where he could'.

Historical evidence shows that the advances of Nigardsbreen in Jostedalen and Brenndalsbreen in Oldedalen led to the most severe damage, and that which affected Tungøyane was the most tragic. The destruction of Tungøyane took place over a period of about 40 years, when the glacier front of Brenndalsbreen was situated in the mouth of Brenndalen, causing a series of avalanches and floods over farmland.

In 1696 the houses at Tungøyane burned down, but they were rebuilt in the same place. From 1702 onwards the Tungøyane farm was regularly damaged by floods and snow avalanches, and the farmers and their families had to move out of their houses during the worst avalanche periods. In 1723 it was stated that the farm was easy to run, but it was situated in front of an advancing glacier (Brenndalsbreen).

During a tax inspection on 12 October 1728, the court stated that the two farmers, before the tax reduction in 1702, paid their taxes according to the instructions from the King and the Church. However, in the late 1720s, the farmers were not able to pay their taxes due to severe damage. Brenndalen, the valley above the farm occupied by the advancing glacier, had previously been good pasture land for cattle. In addition, the farmland around the houses was regularly covered by boulders, sand and gravel from river floods. In 1728 they therefore had to move the houses away from the river plain to a place where they felt safe.

In the middle of summer 1733 the farmland once again suffered severe damage by floods from the glacier. On 2 November 1734 the court (seven persons), led by U. Kås, visited the farm to estimate the damage. At that time, the glacier tongue had advanced through a narrow canyon just above the buildings. The glacier 'that never will disappear', they stated, had advanced down into the main valley. At that time the main river in Oldedalen also changed its course, running over what previously

had been their best farmland. In 1733 the two rivers, together with ice blocks, stones and gravel, covered all the farmland. The farmers were forced to beg for food in order to survive and they were therefore totally unable to pay their taxes. The court found only miserable conditions: starving people and fields covered by ice blocks, boulders, stones and gravel. The court therefore decided (later confirmed by the authorities) that the farmers should not pay taxes for the years 1734–35. When the court visited the farm in 1743, it stated that the glacier tongue was only 60 m from the place where the houses were located before 1728.

On 12 December 1743, an avalanche from the glacier hit the farmhouses rebuilt in 1728. All the houses, people and domestic animals were swept away. After this tragedy the farm was never rebuilt and it was deleted from the land register. Only 80 years earlier, Tungøyane had been one of the wealthiest farms in Oldedalen.

From historical documents, it is possible to reconstruct the natural processes that led to the catastrophe of Tungøyane. Before 1650 they 'saw the glacier as a white cow on the skyline', meaning that there was glacier ice only on the Jostedalsbreen plateau above Brenndalen at that time. In the 1680s and 1690s, the regenerated glacier started to damage the pastures in Brenndalen and caused floods over the farmland in Oldedalen. Around 1700 the glacier front reached the valley mouth above the houses. This means that the glacier advanced 4.5 km in only 50 years (90 m per year on average). Between 1700 and 1728, the glacier flowed through the canyon behind the houses, which were moved in 1728. The 1743 avalanche (ice blocks, water, sand and gravel) from the glacier front, resting on the rock bar above the houses, led to the final destruction of the farm. Length variations of three outlet valley glaciers from Jostedalsbreen (Briksdalsbreen, Stegaholbreen and Fåbergstølsbreen) between 1901 and 1999 are shown in Fig. 5.14.

5.8.2.2 Jotunheimen

Several of the glaciers in Jotunheimen discharge into the Bøvra river, which caused

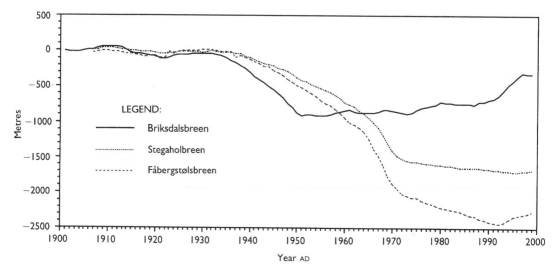

FIGURE 5.14 Length variations of three outlet valley glaciers from Jostedalsbreen (Briksdalsbreen, Stegaholbreen and Fåbergstølsbreen) between 1901 and 1999. (Data: NVE)

severe damage by floods in 1708, 1743, 1760 and 1763. These floods could have been related to glacial activity. It is, however, necessary to interpret these floods with care, unless there is good evidence available. The greatest flood, 'Storofsen', in AD 1789 caused severe damage in eastern Norway and was primarily caused by heavy rain in July.

In Jotunheimen there are no historical documents relating to the Little Ice Age maximum, but lichenometric studies (Matthews, 1977; Erikstad and Sollid, 1986; Matthews and Caseldine, 1987; McCarroll, 1989) and Schmidt hammer 'R-values' (Matthews and Shakesby, 1984) are consistent with the mid-eighteenth century maximum recorded at Jostedalsbreen.

5.8.2.3 Hardangerjøkulen

The maximum Little Ice Age position of Blåisen and Midtdalsbreen, outlet glaciers from Hardangerjøkulen, was around AD 1750 based on lichenometry (Andersen and Sollid, 1971). Calculations of the modern and Little Ice Age ELA on Hardangerjøkulen suggest an ELA depression of ca. 130 m during the Little Ice Age maximum (Nesje and Dahl, 1991a).

5.8.2.4 Folgefonna

At present, Folgefonna consists of three separate glaciers: Nordre, Midtre and Søndre Folgefonna. The two outlet valley glaciers from Søndre Folgefonna, Buarbreen in the east and Bondhusbreen in the west, are the glaciers with the best historical records. A document from 1677 deals with an advance of Buarbreen immediately before that year. A court found severe damage on the Buar farm in 1677 because of rock avalanche and river damage. Many farms surrounding Folgefonna reported damage in the late seventeenth century and early eighteenth century. Bondhusbreen was advancing in the early nineteenth century, and both Bondhusbreen and Buarbreen reached their maximum Little Ice Age positions in the late nineteenth century (ca. AD 1890), while Blomstølskardbreen, a southern outlet glacier from Folgefonna, reached its maximum position around 1940 (Tvede, 1972, 1973; Tvede and Liestøl, 1977).

5.8.2.5 Svartisen/Okstindan

Svartisen, the second largest glacier in Norway, is divided into two parts by the N–S-oriented

Vesterdalen. Beneath the outermost moraine of Fingerbreen, an eastern glacier of Svartisen, the top 2 cm of a peat deposit yielded a radiocarbon age of $695 \pm 75\,yr\,BP$, which is regarded as a maximum age for the moraine (Karlén, 1979). Another radiocarbon date of $600 \pm 100\,yr\,BP$ was obtained from the upper 2 cm of a peat layer under the foreset bed of a delta, which was deposited in a lake dammed by the expansion of a glacier into Glomdalen in the western part of Svartisen. Since these peat layers may have been formed over a considerable time span, the dates cannot be used as precise maximum dates for the ice advance.

In 1800, the front of Engabreen, a SW outlet glacier from Svartisen, was about 30 m from the outer moraine, but it was so close to the sea that it was reached by the sea at flood tides. In 1881, the glacier was 1 km from the fjord. In 1903, however, Engabreen started a minor frontal advance. Later on, annual frontal measurements of Engabreen and Fondalsbreen showed a significant glacier retreat (1–1.5 km) in the 1930s and 1940s (Grove, 1988).

For the Okstindan glaciers, historical records confirm that the last major advance occurred during the first two decades of the tenth century. At this time, some glaciers reached their Neoglacial maximum. The maximum of the Little Ice Age glaciers at Okstindan is represented by moraines with *Rhizocarpon* spp. lichens up to 60 mm in diameter.

However, Karlén (1979) suggested, by means of lichenometry on moraine ridges in the Svartisen, Okstindan and Saltfjellet areas, that glaciers reached their maximum Neoglacial positions prior to the eighteenth century. Innes (1984) suggested that Karlén had underestimated the lichen growth rate in the Svartisen area, and thereby overestimated the moraine ages.

5.8.2.6 Lyngshalvøya

Moraines from the oldest Little Ice Age advance represented in the area occur in front of large glaciers, and lichenometry suggests that they were formed almost contemporaneously. Lichenometry also suggests that the advance took place after AD 1520–1640, while

dendrochronology indicates that this glacier advance occurred before AD 1800 (Ballantyne, 1990). Historical data place the culmination of the readvance in the mid-eighteenth century. Lichenometric, dendrochronological and historical data suggest that the most recent advance culminated in AD 1910–1920, and at a few high-level sites in AD 1920–1930. This advance represents the maximum Neoglacial extent for small glaciers (<ca. $2\,km^2$) in this region.

5.8.2.7 Northern Sweden

The Little Ice Age in northern Sweden comprised several advances, with different glaciers reaching their maximum extent at different times, most of them in the early and late 1600s and in the early and late 1700s. Initially, the climate deteriorated in the 1400s with a period of cold summers between AD 1350–1400. This is evident from lacustrine sediments and moraine formation. In northern Sweden the Little Ice Age is inferred to have begun at approximately AD 1580, with extensive glacier advances between 1600 and 1640 (Karlén, 1976). Smaller maxima are dated at 1650, 1700–1720 and 1810. Subsequent to ca. 1750, glaciers have been retreating, however, with small readvances at approximately AD 1780, 1810, 1820, 1840, 1850, 1870, 1890, 1910 and 1930. During the twentieth century, glacier fronts in northern Sweden have experienced a significant retreat (Holmlund, 1997). Increased winter precipitation during the 1990s, however, has caused positive mass balance on glaciers in northern Sweden. Figure 5.15 is a summary diagram of Little Ice Age glacier variations in Scandinavia compiled from various sources by Boulton *et al.* (1997).

5.8.3 France

There is evidence that the Brenva glacier may have advanced some time after AD 1300 (Grove, 1988). The onset of the Little Ice Age, however, took place in the time interval between AD 1580 and 1645 (Grove, 1988). Cultivated land and forests were covered by ice and

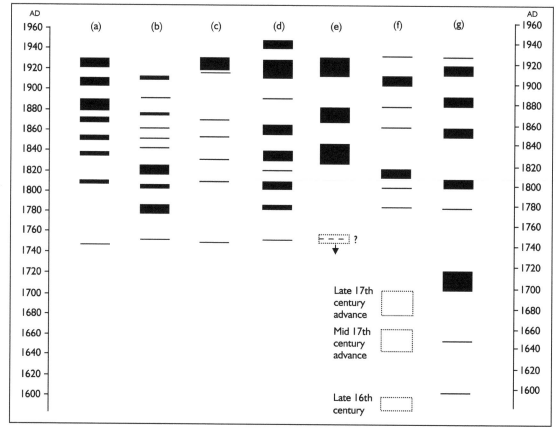

FIGURE 5.15 Little Ice Age glacier advances in Scandinavia: (a) southern Norway; (b) four outlet glaciers from Jostedalsbreen; (c) Storbreen in Jotunheimen; (d) glaciers in SW Norway; (e) Lyngshalvøya in northern Norway; (f) Svartisen in northern Norway; (g) northern Sweden. (Modified from Boulton *et al.*, 1997)

floods as a direct result of glacier expansion. Crops failed in the surrounding valleys. As in western Norway, a series of supplications for tax relief were made. The initial period of glacial advance was followed by a period of less advancing, but extensive glaciers. Minor retreats were followed by advances. These advances, however, caused less damage because the glacier advanced over already spoilt ground. The glaciers in the Mont Blanc Massif do not appear to have advanced significantly in the early eighteenth century. Between about 1750 and 1850, however, the glaciers advanced. The last three significant advances of the Little Ice Age culminated between 1770 and 1780, around 1818–1820,

and around 1850. The glacial maxima ranged between 1835 and 1855. From the mid-nineteenth century to the present, the glaciers in the Mont Blanc Massif have experienced a net recession, despite several advance phases (Fig. 5.16).

5.8.4 The Alps

The most recent interval of glacier advance occurred in the six centuries between about AD 1250/1300 and AD 1850/1860 (Grove, 1997), during which some outlet glaciers extended 2–2.5 km beyond their present marginal positions. The main Little Ice Age glacial advances in the Alps occurred around

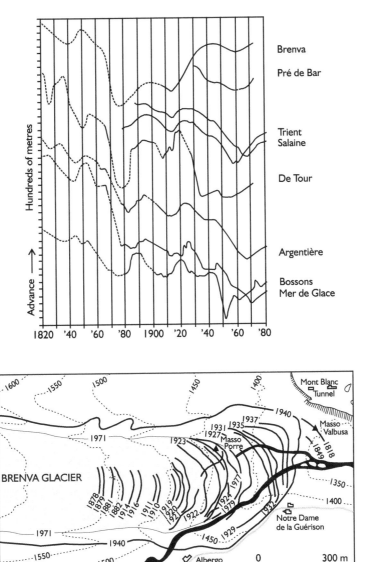

FIGURE 5.16 Fluctuations of the major glaciers in Mont Blanc since AD 1820 (top) and frontal positions of the Brenva Glacier between 1818 and 1979 (bottom). (Adapted from Grove, 1988)

AD 1350, 1600–1650, 1770–1780, 1815–1820 and 1850–1860.

From the records of glacier front variations, mass-balance reconstructions, temperature and precipitation data, Kuhn *et al.* (1997) concluded that glacier activity since 1860 has been generally homogeneous in the Alps. There was a short period at the end of the nineteenth century when regional variability of precipita-

tion may have caused different accumulation. During the last two decades of that century, glaciers had nearly reached equilibrium size after a rapid decrease following their mid-century maxima. After the 1920 advance period, Alpine glaciers were not as close to equilibrium as before and during the period 1965 to 1985. The 1930–64 period was characterized by higher continentality, strong retreat, and rather uniform response of all Alpine glaciers.

The climate in the European Alps during the twentieth century has been characterized by an increase in minimum temperatures of approximately 2°C, a smaller increase in maximum temperatures, and a decrease in sunshine duration through to the mid-1980s. The temperature increase was most pronounced in the 1940s and 1980s. Since the mid-1850s (peak of the Little Ice Age) the glaciated area has been reduced by 30–40 per cent, and by about half of the glacier volume (Haeberli and Beniston, 1998).

Wurtenkees (about 1 km^2) in the Eastern Alps has been one of the most strongly retreating glaciers in this region (Schöner *et al.*, 1997). Under present climatic conditions the glacier needs a summer temperature depression of 1–1.5°C to return to a balanced mass budget. Under temperature scenarios predicting future global warming, the glacier will probably disappear in the first part of the next century.

5.8.4.1 Ötztal, Eastern Alps

The largest group of glaciers in the Eastern Alps is located at the head of Ötztal, where mountains on the Italian border rise to more than 3600 m. When Vernagtferner advances across the floor of Rofental, it obstructs the flow of rivers draining the glacier upvalley. It has been suggested that the Vernagtferner may be a surging glacier, at least periodically. When the glacier advances down into the valley bottom, lakes form, and if the damming phase is sufficiently long, it can cause severe flood damage further downstream after violent overspills. Government inquiries are the main sources of information about the glacier fluctuations in the early part of the Little Ice Age. The onset of the Little Ice Age advance of Vernagtverner occurred during the time period 1599–1601 (Fig. 5.17). During the period 1678 to 1725, the Ötztal glaciers advanced into the main valley bottoms, damming the river upstream. The subsequent floods caused severe damage downstream. In the 1770s both the Vernagt and Gurgler glaciers were advancing; however, between 1822 and 1840 the Vernagtferner retreated considerably. During the time interval between 1845 to 1850, on the other hand, the glaciers in the Ötztal advanced. Since 1848, there have been no floods from the Vernagt lake. This lake was first formed in 1599, emptied rapidly on 25 July 1600, and emptied again slowly in 1601. The second phase of lake formation was in 1678. The lake emptied rapidly on 16 July 1678 and 14 June 1680, while the lake emptied slowly in 1679 and 1681. The third phase of lake formation took place in 1771, with slow emptying in 1772 and in 1774 and a rapid emptying on 23 July 1773. The last phase of lake formation occurred in 1845. The lake emptied slowly in 1846, but on 14 June 1845, 28 May 1847 and 13 June 1848 the lake emptied rapidly (e.g. Grove, 1988). The glaciers in the Ötztal Alps continued to retreat from 1850 until about 1964, interrupted by minor advances between 1890 and 1900 and around 1920. The changes in the areal extent of Hintereisferner and Kesselwandferner since 1847 are shown in Fig. 5.18, p. 155. A comparison of the glacier fluctuations in the Mont Blanc massif and in the Ötztal region shows that the advances and retreats were closely in phase (e.g. Grove, 1988).

5.8.4.2 Italian Alps

Most Italian Alpine glaciers reached their Little Ice Age maximum extent around AD 1820, when glaciers extended up to 2 km beyond their present position (Orombelli and Mason, 1997). The second largest glacier advance in 1845–1860 occurred subsequent to a retreat in the 1830s. The glaciers retreated again up to 1 km upvalley from their maximum position by around 1870. In the 1880s, however, glaciers readvanced to reach a less extensive maximum position by around 1890/1895. Some glaciers reached their last maximum around 1920–1925, after which glaciers experienced a long, continuous retreat from the 1930s to the 1950s. During the 1960s, 1970s and parts of the 1980s, glaciers have advanced, but in the 1990s glaciers have been retreating.

In the Lombard Alps in the central sector of the Italian Alps, all glaciers have been retreating

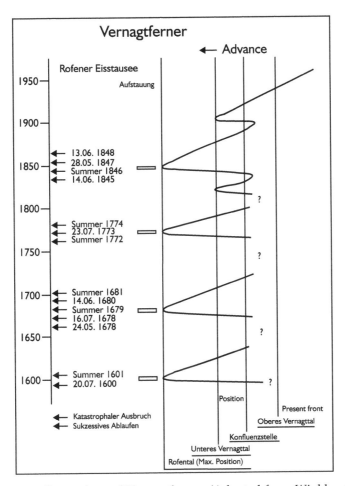

FIGURE 5.17 Little Ice Age fluctuations of Vernagtferner. (Adapted from Winkler, 1996)

since the beginning of the twentieth century; however, the trend has not been uniform (Pelfini and Smiraglia, 1997). Since the 1950s, there has been a drop in the number of retreating glaciers and an increase in stationary and advancing glacier termini. A new recession period started in 1985. The glacier fluctuations in the Lombard Alps are well correlated with temperature records from the region, with a response time of approximately 20 years.

5.8.4.3 Switzerland

There is strong evidence of advancing glaciers before the sixteenth century from the eastern part of Valais and from the

Bernese Oberland. Tree logs from within, and soils from beneath, moraine sequences have been radiocarbon dated, the majority of them giving dates ranging from the eighth to the tenth century. Investigations indicate that the glaciers advanced after AD 1100 and before the sixteenth century. The glacier fluctuations of the Unterer Grindelwaldgletscher are shown in Fig. 5.19. The glacier was more extensive between 1600 and 1870 than it has been since. The Unterer Grindelwaldgletscher reached its maximum extent between 1590 and 1640. Regular measurements of the frontal position of the Grindelwaldgletscher began in 1880.

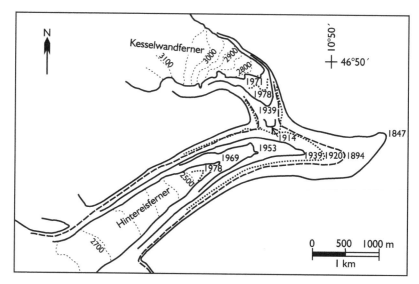

FIGURE 5.18 Changes in the extent of Hintereisferner and Kesselwandferner since 1847. (Modified from Grove, 1988)

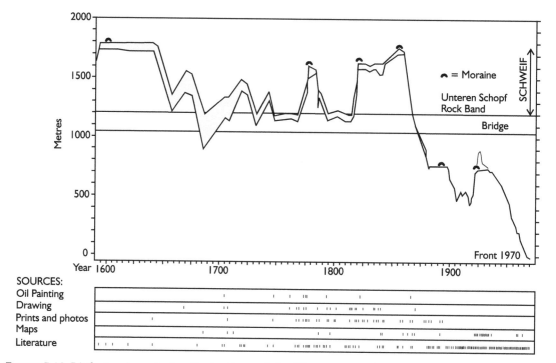

FIGURE 5.19 Little Ice Age glacier fluctuations of the Unterer Grindelwaldgletscher. (Adapted from Grove, 1988)

5.8.5 Eurasia

Glaciers in the mid-latitude mountains of Eurasia have retreated significantly during the last century. Measured and reconstructed glacier mass balances show that glacier retreat began around the 1880s. The mean annual mass-balance value for 1880–1990 has been estimated at −480 mm for glaciers under maritime influence, and −140 mm for continental glaciers (Mikhalenko, 1997).

During the Medieval period, the ELAs in the Caucasus were higher and the glaciers less extensive than at present. Glaciers advanced between the thirteenth and fifteenth centuries, between 1640 and 1680, and between 1780 and 1830. Around the mid-twentieth century the glaciers stopped retreating. In the early 1960s the ELA was lowered by 200–300 m compared with the previous decade, as a result of lowered summer temperatures and increased winter precipitation. A further increase in precipitation during the 1960s caused further glacier expansion. By 1979, however, only six of 26 glaciers were still advancing.

Himalayan glaciers have predominately retreated since 1880. Moraines, however, indicate that the retreat has not been continuous. In this region, temperature is the most critical factor for the glacier mass balance, since temperature determines whether the monsoon precipitation falls as rain or snow. The Karakoram glaciers retreated from advanced positions in the mid- to late nineteenth century. In the 1890s and the first decade of the twentieth century, however, they advanced as a result of intensified monsoonal airflow (Grove, 1988, and references therein). Between 1920 and 1940 the majority of the glaciers were either stationary or advancing. After 1940 glaciers mainly retreated, but the glacial retreat halted or reversed in the 1970s.

The patterns of retreat from maximum Little Ice Age positions to the present were studied by Savoskul (1997) at 20 glaciers in the relatively humid northwestern front ranges and arid inner areas of Tien Shan, central Asia. She found that the large Little Ice Age glaciers in the warm and humid northwestern mountain ranges were 1.5–1.9 times larger than the modern glaciers. The Little Ice Age glaciers in the cold and arid inner parts of Tien Shan were only 1.03–1.07 times larger. The maximum Little Ice Age ELA depression was 100–200 m in the humid areas and 20–50 m in the arid areas.

5.8.6 China

In China, the major glacierized areas are in the northwest: in the Tien Shan, Kunlun Shan, and the Himalayas. Between the mid-nineteenth and mid-twentieth centuries, the glaciers retreated as a result of temperature rise. The warmest five years in the 1940s were 0.5–1°C warmer than the mean of the last 100 years. The termini of the longest glaciers retreated several hundred metres to several kilometres. Between the mid-1950s to the mid-1970s, 22 glaciers studied in the Qilian Shan were retreating, some of them more than 20 metres a year. In the interior of Tibet, however, the retreat was less extensive. During recent decades, the glacial retreat has slowed down (e.g. Grove, 1988).

5.8.7 North America

In the Cascade and Olympics mountains, the South Cascade glacier reached its maximum Little Ice Age position in the sixteenth or seventeenth century, while the outer moraines of the Le Conte and the Dana glaciers date to the sixteenth century. The glacial advances during the sixteenth and seventeenth centuries were followed by retreat, and again by minor advances during the nineteenth century. Mt. Mazama ash helps to date the moraine sequences.

In the Canadian Rockies, major advances seem to have occurred in the late seventeenth to early eighteenth, early to mid-nineteenth, and late nineteenth to early twentieth centuries. A compilation of glacier fluctuations in the Canadian Rockies (Grove, 1988) shows a significant glacier retreat, especially after 1910–20 (Fig. 5.20). Around 1945 there was a change from significant glacial recession towards stability or even advance.

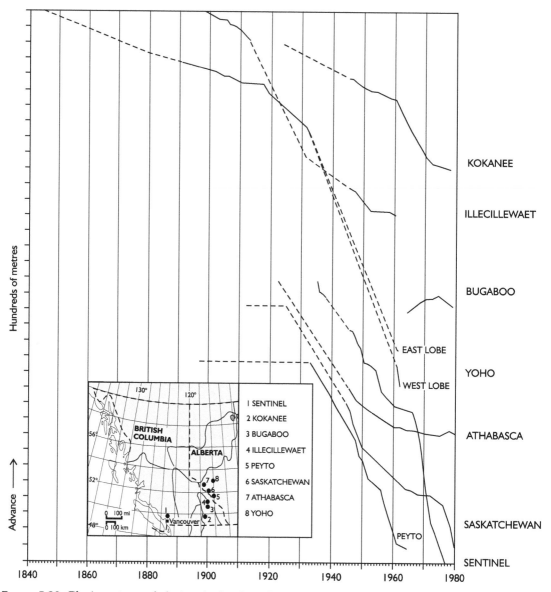

FIGURE 5.20 Glacier retreat of glaciers in the Canadian Rockies between 1840 and 1980. (Adapted from Grove, 1988)

Glaciers in Alaska display moraine evidence of Little Ice Age cooling peaks in the seventeenth and nineteenth centuries. Recent evidence from glaciers on the Seward Peninsula suggest that the ELA fell by about 170 m at that time (Calkin *et al.*, 1998).

Palaeoclimate records from lake sediments, trees, glaciers and marine sediments have been compiled to provide an insight into environmental change over the last four centuries in the circum-Arctic region (Overpeck *et al.*, 1997). Between 1840 and ca. 1950 the Arctic warmed to the highest temperatures recorded during the last four centuries, terminating the Little Ice Age. This warming has led to glacier retreat, melting of permafrost and sea ice, and a change in terrestrial and lake ecosystems. The cause of this warming is probably related to an increase in atmospheric trace gases, increased solar radiation, decreased volcanic activity, and internal climate feedback mechanisms.

5.8.8 South America

Maps and photographs demonstrate that the glaciers in Venezuela have retreated significantly during the twentieth century. Some glaciers have thinned by 100–150 m and the ice-covered area has been reduced by as much as 80 per cent (Grove, 1988). A rise in the ELA and glacier retreat in the tropical Cordillera Blanca may be explained by a combination of a spatially uniform rise in air temperature and a decrease in humidity, with geographically different effects (Kaser and Georges, 1997).

The historical fluctuations of Gualas and Reicher Glaciers on the North Patagonian Icefield, southern Chile, have been dated by dendrochronology (Harrison and Winchester, 1998). Vegetation trimlines were dated to AD 1876, 1909 and 1954. Intermediate stages of recession of the Gualas and Reicher glaciers were dated to the early 1920s, mid-1930s, and 1960s. The glacier fluctuations were interpreted to reflect fluctuations in winter precipitation rather than summer temperatures.

5.8.9 Greenland

The most extensive data on the behaviour of local glaciers beyond the ice sheet come from Sukkertoppen and Disko Island. Similar to the inland ice-sheet lobes, the majority of the local glaciers reached their maximum Neoglacial extent before the eighteenth century, possibly as early as AD 1750. Glaciers started to retreat around 1850, but between 1880 and 1890 the glaciers were reactivated, causing glacier advances. In the early twentieth century, the glacier recession continued, however, interrupted by some advance periods. The fastest glacial retreat took place between the 1920s and 1940s. A striking feature of the Little Ice Age record from Greenland is the synchroneity of the fluctuations of the continental ice sheet and the local glaciers (Gordon, 1980).

5.8.10 Africa

Moraine sequences in front of the glaciers demonstrate more advanced positions during previous centuries than at present. Between 1899 and 1974 the area of the Lewis glacier on Mt. Kenya was reduced from $0.63\,\mathrm{km}^3$ to $0.31\,\mathrm{km}^3$ and the elevation of the front rose by 130 m. Glacier melting was extensive in the early part of the century, but slowed down from the early 1930s to the early 1960s, after which the terminus has continued to retreat. The shrinkage of the East African glaciers during the last century seems to be a combined effect of reduced precipitation with accompanied reduced cloudiness, and increased temperature.

5.8.11 The Pyrenees

The Little Ice Age glacier history of the Pyrenees has been reconstructed from documentary, cartographic and photographic evidence (Grove and Gellatly, 1997). All glaciers seem to have expanded during the Little Ice Age, and some fronts advanced more than 750 m and descended almost 200 m in elevation. The glaciers were in advanced positions during the late eighteenth and mid-nineteenth centuries. In the 1860s and 1870s glaciers retreated significantly. During the twentieth century, glaciers have advanced in the 1900s, 1920s, 1940s, 1960s, and late 1970s.

5.9 Glaciers, environmental change and the human race

Glacier chronologies obtained from different parts of the world have demonstrated that glaciers respond in different ways to climatic triggers. Predictions of future climate trends and glacier response must therefore avoid general statements about the linkage between recent glacier variations and climate change. Global warming scenarios invoke higher sea-levels, increased melting of the glaciers, and accelerated calving rates for the West Antarctic ice sheet. A warmer climate does not, however, necessarily mean that all the glaciers will disappear, because some regions may experience increased precipitation during the accumulation season.

The impact of humans on the environment depends not only on the nature of human society, but also on the nature of the environment (e.g. Kemp, 1994). As there are, in fact, many 'environments', there are many possible responses to human interaction. The different environments vary in scale and complexity, but they are closely linked, and in combination constitute the earth–atmosphere system, comprising a series of interconnected components or subsystems ranging in scale from microscopic to continental. In the past, human environmental impact was mainly at the subsystem level, but, the growing scale of interference has caused the impact to extend to a continental or even global scale.

5.10 Models of Late Quaternary climate and ice-sheet evolution

At present, glaciers and ice sheets cover about 10 per cent of the Earth's surface, locking up about 33 million km^3 of freshwater, which would correspond to a sea-level rise of 70 m. During the Quaternary glacial periods, ice sheets were more extensive, covering about 30 per cent of the Earth's land area. Glaciers and ice sheets are also sensitive to climate change, varying in response to winter precipitation and summer temperature. To answer the questions of why, where and when glaciers form, it may be helpful to recognize glaciers as systems, with inputs, outputs, and interactions with other systems. Mass and energy enter the system in the form of precipitation, gravity, solar radiation and geothermal heat, and leave the system in the form of vapour, water, ice, debris and heat. The mass and energy is transferred through the system at various speeds and periods of storage. The most important input to glacier systems is, however, snowfall, windblown snow and snow avalanches from adjacent valley sides. Glaciers grow where climatic and topographical conditions allow the input to be greater

than the output (accumulation > ablation) and vice versa.

Due to the growing evidence for global climate change and ice-sheet variations, and as a result of the complexity of the different records, scientists are turning to simulation modelling for explanations of the causes of observed changes. Conceptual models aim to highlight the connections and feedbacks between different components of the system. These models can later be tested empirically (by observation) and by simulation modelling. Imbrie *et al.* (1992) developed a conceptual model to examine the response of Atlantic Ocean circulation to solar radiation changes. They proposed four circulation modes:

(1) During an interglacial, deep convection in the Atlantic is concentrated into three cells: in the Nordic Seas, in the open ocean, and in the Antarctic seas.
(2) During a pre-glacial episode following an interglacial, the Nordic heat pump is reduced in strength or stopped, while the Antarctic circulation cell is enhanced.
(3) During a full glacial, the open ocean heat pump operates at a maximum, while the Antarctic cell is stronger than during an interglacial.
(4) During deglaciation, the Nordic cell is restabilized, while the open ocean heat pump is stronger than during interglacial periods.

This model demonstrates how a change in one oceanic circulation component immediately affects the entire system. Reduced terrestrial radiation receipts as a result of Milankovitch forcing causes glacial ice cover and sea ice to expand. During this stage, the Nordic heat pump becomes gradually reduced and flow is concentrated in the open ocean. Evaporation from the ocean surface and lowering of the temperatures will occur simultaneously with a shift in the position of the westerly winds and more enhanced ocean circulation in lower latitude oceans. In turn this may effect the strength and location of the monsoon cells and the position of the intertropical convergence zone. Terrestrial and marine Quaternary proxy data support the evidence that changes in the North Atlantic circulation influence climatic changes in other parts of the world (e.g. Lowe and Walker, 1997).

Kukla and Gavin (1992) presented a model of how the Earth may respond to insolation variations as a result of Milankovitch forcing. Their model indicates that the three insolation components in combination drive the Earth's climate into an ice age. The main reasons for this are precessional changes (gradual decrease in insolation between July and November), reduced receipt of radiation in high latitudes as a result of obliquity variations, and reversed seasonal insolation variations in lower latitudes, leading to increased insolation during the spring. Their model suggests a gradual decrease in radiation, but feedback processes force the system into more rapid cooling. The cooling in higher latitudes combined with increased northern transport of water vapour causes build-up of glaciers. In their model, orbitally induced changes magnify short-term effects (such as changes in solar activity and volcanic eruptions) when they are in phase, and damp them when they are out of phase.

Oeschger (1992) presented a hypothesis for Late Quaternary climate change based on the interaction between changes in ocean thermohaline circulation and variations in atmospheric gases. His hypothesis included a change between a glacial state with reduced North Atlantic Deep Water (NADW) formation, and warmer climates when NADW formation was strengthened, sea-level was higher, and atmospheric concentrations of CH_4 and CO_2 were at present levels. In Oeschger's model, changes in the distribution of solar irradiance cause changes in NADW formation. In addition, changes in ocean chemistry and the operation of the biological pump as a result of sea-level fluctuations cause variations in the CO_2 flux between the ocean and atmosphere. Atmospheric CO_2 concentrations may be buffered by a net flux from terrestrial biomass because of reduced vegeta-

tion cover during glacials. This may explain why the decline in atmospheric CO_2 content lagged behind the lowering of the global temperature at the end of the Eemian interglacial. The reduction in CH_4, however, coincided with the climatic deterioration as a result of increased aridity and reduced areas of wetland, and therefore reduced CH_4 flux to the atmosphere.

During glacial periods, NADW formation is reduced. Changes in critical boundary conditions like solar irradiance, sea-level change, and the extent of ice cover (albedo) may cause rapid switches in NADW formation. The Dansgaard–Oeschger cycles have been attributed to a combination of solar irradiance and damped effects of NADW formation. Changes in NADW formation, and thereby the biological heat pump in the oceans, have led to changes in the atmospheric CO_2 concentration of between approximately 200 (reduced NADW formation) and 240 ppm (enhanced NADW formation). Enhanced NADW formation at the beginning of the Holocene increased the heat transport to northern latitudes and increased melting of the northern ice sheets. The atmospheric CO_2 content paralleled the global sea-level rise.

Variation in insolation induced by Milankovitch forcing is considered to be the main driving force for glacial–interglacial cycles. Superimposed on these cycles are, however, a series of complex internal feedback mechanisms, such as ocean heat transfer, albedo and gas exchange. How these different components interact to cause climatic changes is, however, not well known. Of special interest are the complex relationships between atmospheric gas exchange, ocean thermohaline variations, variations in meltwater flux, and ice-sheet fluctuations. It is often difficult to establish precisely the order of events as a result of poor stratigraphic resolution and lack of precise dating and correlation. It is therefore difficult to solve the 'chicken and egg' dilemma.

6

Late Quaternary sea-level changes

6.0 Chapter summary

The chapter explains the principles of glacio-eustasy and glacio-isostasy and causes of relative sea-level changes. The chapter also presents evidence of recent changes in sea ice extent. Finally, models of future sea-level changes are presented. The chapter stresses the significance of glaciers for sea-level change, the significance of sea-level change for glaciers, and the sea ice feedback effect on climate change at high latitudes.

6.1 Glacio-eustasy and glacio-isostasy

The basic idea of *isostasy* is that the Earth's crust, with a mean density of approximately 2800 kg m^{-3}, is floating on the underlying plastic mantle with a mean density of about 3300 kg m^{-3}. The amount of *crustal depression* resulting from ice sheet loading is a function of ice thickness and the ratio between the densities of ice and rock. The density of ice is about a third that of the crust, and therefore the crustal depression beneath an ice sheet is about a third of the ice thickness. Normally, the amount of crustal depression increases from the margin towards the centre of the ice sheet, where the ice sheets in most cases are thicker. The marginal depression, however, continues up to 150–180 km beyond the margin of the ice sheets. Therefore the sea can transgress proglacial areas.

The growth and melting of glaciers and ice sheets have a significant effect on global sea-level, causing large regional and global sea-level changes during glacial–interglacial cycles. A relative rise (*transgression*) or fall (*regression*) in sea-level caused by glaciers and ice sheets can occur by *glacio-eustasy, glacio-isostasy, hydro-isostasy* or *geoidal eustasy*.

Superimposed on long-term trends in sea-level change caused by tectonic activity, changes in mass distribution and the shape of the Earth, changes in the volume and mass of the hydrosphere, and the effects of variations in the rate of rotation or in the axis of tilting of the Earth, are major sea-level oscillations due to expansion and contraction of the ice sheets. During expansion of terrestrial ice sheets, water is extracted from the oceans. During the last glacial maximum, the terrestrial volume of the ice sheets produced a *glacio-eustatic* (controlled by the growth and contraction of ice sheets) sea-level lowering of about 120–130 m. In contrast, melting of the Greenland and Antarctic ice sheets would cause sea-level rises of 5.5 and 60 m, respectively (total glacio-eustatic sea-level rise of about 70 m). In some of the tectonically stable areas of the world, the glacio-eustatic sea-level variations have been reconstructed. In Bermuda (a stable mid-oceanic carbonate platform), uranium-series and amino-acid dating of corals and speleothems from fossil coral reefs and beach deposits have been used to reconstruct eustatic sea-level changes during the last 250,000 years

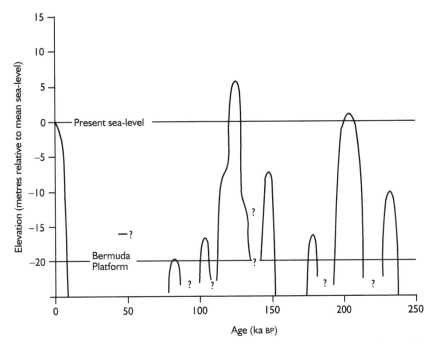

FIGURE 6.1 The Bermudan record of Late Pleistocene sea-level fluctuations. (Adapted from Harmon *et al.*, 1983)

(Fig. 6.1). According to the Bermuda record, only on two occasions during that period has sea-level been higher than at present; at approximately 200,000 yr BP (+2 m) and during the last Eemian interglacial (+5–6 m).

Due to the limitations of local sea-level records, oxygen isotope records from deep ocean sediments have been used to reconstruct sea-level changes. The oxygen isotope record provides a proxy for ice-sheet volume and glacio-eustatic sea-level change (e.g. Shackleton, 1987). Although the curves of oxygen isotope variations through time reflect ice sheet and ocean volume changes, however, absolute changes in water depth are difficult to calculate.

Glacio-eustatic sea-level changes are most reliably recorded for the last 13,000–15,000 yr BP, when the Laurentide, Fennoscandian and British ice sheets retreated and finally melted. In addition, the Antarctic and Greenland ice sheets decreased in volume. A record based on high-precision U–Th dating of fossil corals on Barbados (Fig. 6.2) shows that sea-level was around 121 ± 5 m lower than at present during the last glacial maximum, and global sea-level

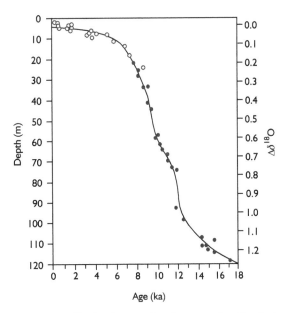

FIGURE 6.2 Barbados sea-level curve for the last 17,000 years based on U–Th dating of submerged corals. (Adapted from Fairbanks, 1989)

rose to around $-60\,\mathrm{m}$ at $10,000\,\mathrm{yr}$ BP (Fairbanks, 1989). The Late-glacial sea-level rise was interrupted by two major meltwater pulses, at 14,000 and $11,000\,\mathrm{yr}$ BP (Bard *et al.*, 1990).

Glacio-isostasy is crustal deformation resulting from the build-up and decay of great ice sheets. The crustal deformation varies with the rigidity of the crust. The depression at one place must be compensated elsewhere, and hence marginal displacement of the crust involving upward bulging (forebulge) may be one aspect of this compensation (Peltier, 1987). The distance between the margin of the ice sheet to the fore-bulge depends on the flexural parameter of the crust, or the amplitude of bending of the litho-sphere, which is mainly related to lithospheric density, thickness and elasticity.

Glacio-isostatic recovery can be considered as a process that accelerates rapidly and then slows gradually. Glacio-isostatic recovery in response to deglaciation can be subdivided into three phases (Andrews, 1970):

(1) Restrained rebound occurs beneath a thin-ning ice sheet. This period is not recorded by direct sea-level data because the area is covered by ice.
(2) Postglacial uplift is the rebound phase after deglaciation. Relative sea-level variations can be recorded by means of geomorpholo-gical and sedimentological evidence.
(3) Residual uplift is the rebound that takes place several thousand years after the region is deglaciated. Some regions that were occupied by ice sheets during the last ice age are still rising at present because of the long response time of litho-spheric recovery. In many areas isostatic uplift is considered not to be complete. Highland Britain still rises by almost $0.2\,\mathrm{cm\,yr^{-1}}$ (Shennan, 1989), the Gulf of Bothnia by $0.8{-}0.9\,\mathrm{cm\,yr^{-1}}$ (Broadbent, 1979), while the eastern Hudson Bay area in Canada experiences an uplift of about $1.1\,\mathrm{cm\,yr^{-1}}$ (Hillaire-Marcel, 1980).

Geophysicists are interested in constraining their models of lithospheric deformation using empirical data on crustal movements, while geomorphologists want to reconstruct dispersal centres of former ice sheets. The complete response of sea-level in an area can be divided into three segments of the sea-level curve: (a) the ice build-up or loading phase which occurs during ice sheet expansion; (b) the equilibrium phase with stable high sea-level when the ice sheet is at maximum thick-ness; and (c) the deglacial unloading phase characterized by relative sea-level fall.

6.2 Relative sea-level changes

A wide range of sediments and landforms may be used to reconstruct sea-level histories, such as deltas, beaches, shore ridges and erosional platforms marking former shorelines. More accurate sea-level data can be obtained from sediment cores retrieved from small lakes and bogs situated at different elevations in a small area or along the same isobase (e.g. Svendsen and Mangerud, 1987).

In northwest Europe, the sea-level rise since deglaciation has been reconstructed at numer-ous sites along the coastline, mainly from estuarine sediments and coastal basins. Coastal lake basins may provide the most accurate his-tory of sea-level changes. As soon as these basins, preferably with a bedrock threshold, have been raised above sea-level, they start to accumulate brackish and freshwater sediments (Fig. 6.3). Such basins are referred to as isola-tion basins (e.g. Svendsen and Mangerud, 1987). A marine transgression may lead to the resubmergence of these basins, and brackish or marine deposits will accumulate above freshwater sediments. It must be emphasized that it is the position of sea-level relative to the outlet of these basins that determines whether the sediments are marine, brackish or lacus-trine. The timing of sea-level being level with the outlet of each lake is determined by identify-ing (using biostratigraphical methods) and dating the boundary between the brackish and lacustrine sediments in the cores. This method, however, requires that there is no hiatus between the brackish and lacustrine sequences. Pollen and diatom biostratigraphy can normally be used to test whether the sedimentation is continuous across this boundary. The elevation and date of former sea-levels as determined for

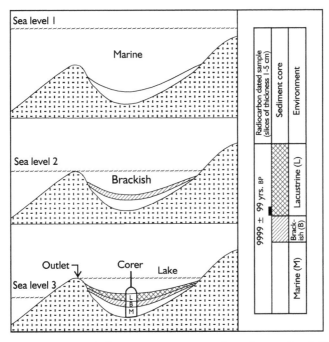

FIGURE 6.3 A bedrock basin at three different stages. Top: when the sea-level (1) was well above the threshold; centre: when the sea-level (2) was at the threshold; and bottom: when the sea-level (3) was below the threshold. A typical core sequence, and the level from which radiocarbon samples are collected, is shown to the right. To construct a sea-level curve, basins from different elevations are cored. (Adapted from Svendsen and Mangerud, 1987)

each locality are plotted in an age–elevation diagram, and the relative sea-level curve is the regression line between the points for each lake. The resolution of the curve depends on how many localities from different elevations are included, and therefore the resolution varies both within each diagram and between different diagrams. To avoid problems with reservoir age and resedimentation, radiocarbon dates should preferentially be taken from the lacustrine/brackish boundary. During a regression phase the radiocarbon age may be slightly younger than the boundary, and during a transgression slightly older.

Along the margins of the Quaternary ice sheets, shorelines formed. As deglaciation and uplift proceeded, lower shorelines were formed. Younger shorelines are tilted less steeply than older shorelines due to a decreasing amount of differential uplift through time. This is clearly seen in an equidistant shoreline diagram. Because the outer fjord areas were

deglaciated earlier than the inner parts, the lower shorelines can normally be traced further inland than the higher ones. This can be observed in Scotland, Scandinavia, the Canadian Arctic, and the eastern USA. Subsequent to deglaciation, shorelines developed as a result of the complex interplay between glacio-isostatic uplift and glacio-eustatic sea-level rise. Where prominent shorelines of the same age are found in different areas, *isobases* can be constructed for these shorelines. Isobases do not show absolute uplift since shoreline formation, but the total amount of uplift minus the glacio-eustatic component of sea-level change. As an example, if a shoreline at an altitude of 40 m is dated to 9000 yr BP, when the global eustatic sea-level was ca. 35 m lower than at present, it indicates a total uplift of 75 m. It is important to remember that isobases join points of equal altitude or uplift of the same age. The isobase pattern can either be reconstructed manually or by means of trend surfaces, and gives a 3D

picture of the crustal deformation from ice loading. Recently, models of glacio-isostatic rebound
and sea-level fluctuations have been developed
(Lambeck, 1991a,b, 1993a,b). These models com-
bine glaciological data, empirical data related to

sea level variations, and geophysical parameters
to produce numerical models that are in good
agreement with the actual sea-level history.
    Clark *et al.* (1978) and Clark (1980) divided
the surface of the Earth into six sea-level

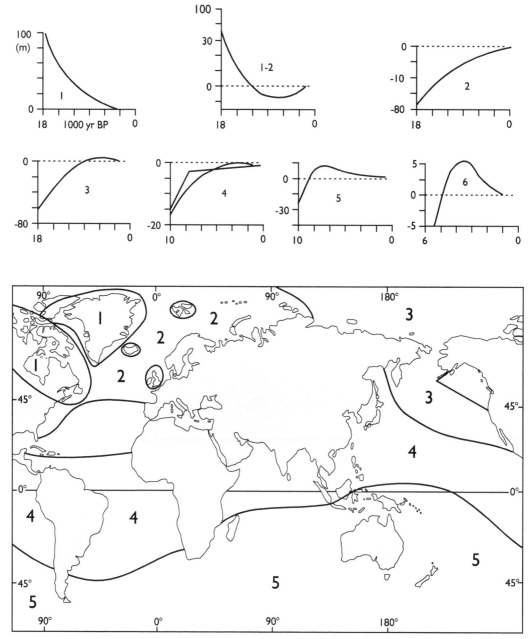

FIGURE 6.4 Distribution of sea-level zones and typical relative sea-level curves. (Adapted from Clark
et al., 1978)

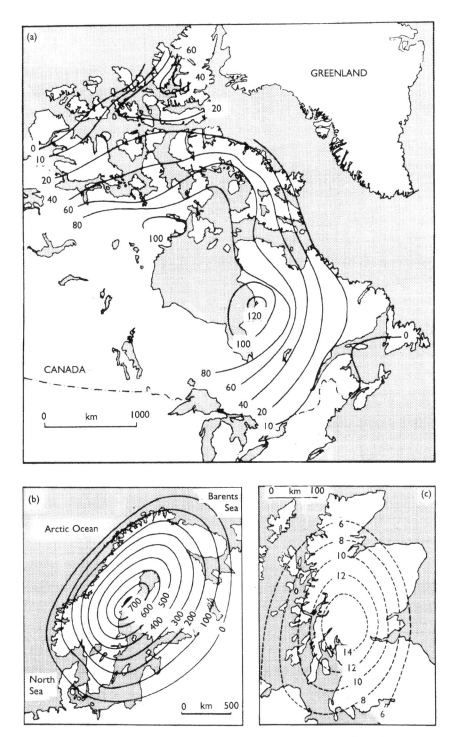

Figure 6.5 Regional isobase maps. (a) Shoreline emergence in eastern Canada since approximately 6000 yr BP. (b) Absolute uplift in Scandinavia during the Holocene. (c) Isobases for the Main Postglacial Shoreline (ca. 7000–6000 yr BP) in Scotland. (Modified from Benn and Evans, 1998)

zones (Fig. 6.4) based on typical postglacial relative sea-level curves. Zone 1 is within the limits of the large Pleistocene ice sheets, characterized by continuous land uplift (regression). Zone 2 occurs beyond the limits of the Pleistocene ice sheets, where the sea-level history is influenced by eustatic submergence modified by forebulge collapse. Zone 3 is further away from the former ice sheets. This zone is characterized by initial eustatic submergence, followed by emergence several thousand years after deglaciation. Zone 4, located in the tropics and subtropics, is characterized by continuous eustatic submergence. In zone 5, in the southern oceans, the sea level is initially controlled by eustatic submergence; however, when glacial meltwater stops draining to the oceans, slight emergence takes place. Zone 6 includes all continental margins except those lying in zone 2, and is characterized by slight emergence after meltwater stopped flowing into the oceans.

Within the regions covered by the Pleistocene ice sheets (zone 1), the change in sea-level has been characterized by isostatic uplift and regression. The pattern of uplift can be studied from isobase maps. Isobase maps provide information on the isostatic loading of the crust and therefore dispersal centres of ice sheets. Generalized isobase maps for eastern Canada, Scandinavia and Scotland are shown in Fig. 6.5. The uplift pattern in Scandinavia and Scotland indicates one main loading centre, while the isobases over eastern Canada show three loading centres, indicating multiple domes on the Laurentide ice sheet. In areas with detailed investigations, isobase maps are in general more complex, exhibiting abrupt discontinuities, reflecting possible fault or fracture zones.

6.3 Sea ice

Iceland is located where the warm water from the Atlantic and the cold water from the Arctic meet. The Irminger current, a branch of the North Atlantic Drift, is deflected westwards by a submarine ridge and flows along the south and west coast before it sinks below the East Greenland current. A branch of the East Greenland current flows around the north and east coast of Iceland and sometimes brings drifting sea ice close to the coast. On some occasions, ice drifts further west along the east coast of Greenland through the Denmark Strait. In a 'normal' sea-ice year, the edge of the ice is about 90–100 km from the northwest coast of Iceland from January to April. In a mild year, the edge is ca. 200–240 km off the coast, in a severe year the ice extends along the northern coast, and in an extreme year the ice may reach the south coast. Koch (1945) reconstructed variations in sea-ice extent since about AD 800 based on a compilation by Thoroddsen. His reconstruction was based on stretches of coast that were 135 km long, and his index was the product of the number of weeks that sea ice was observed and the number of stretches from which sea ice was seen (Fig. 6.6).

A compilation by Bergthórsson (1969) of severe ice years since Norse settlement shows common features with Koch's reconstruction. Due to a strong correlation between air temperature and sea ice, Bergthórsson (1969) reconstructed decadal running means of annual mean temperature variations from the sea-ice data (Fig. 6.7). Sigtryggsson (1972) prepared a more detailed reconstruction of annual variability of sea-ice extent. Based on a critical examination of available data, Ogilvie (1984) made a new reconstruction of sea-ice extent from the Medieval period to AD 1780.

From the documentary climate data available from Iceland, Ogilvie (1998) suggested that colder conditions prevailed during the last decades of the twelfth century, and that there was much sea ice in the seas between Iceland and Greenland around the mid to latter part of that century. The fourteenth century seems to have been very variable, however, with cold periods around 1320 and in the late 1340s. From the 1350s to the end of the 1370s, the climate was generally cold with famine. The sailing route from Iceland to Greenland was changed in the 1350s due to sea ice. Icelandic sources also indicate extensive sea ice some time around the mid to late 1300s. Cold

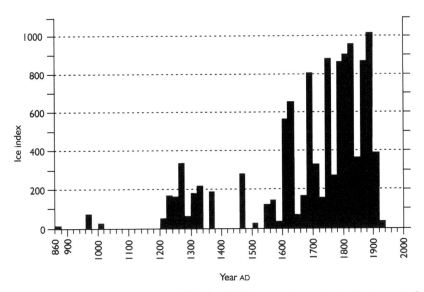

FIGURE 6.6 Sea-ice incidence around Iceland (Koch, 1945) from the ninth to the twentieth century.

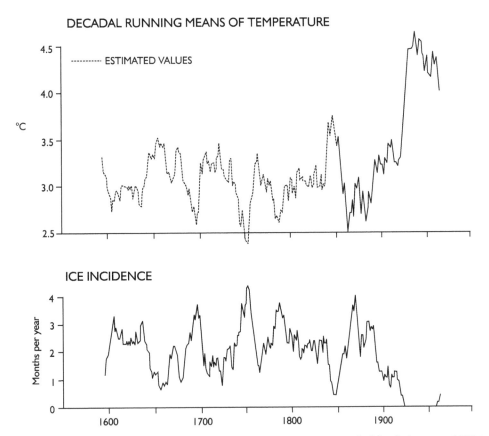

FIGURE 6.7 Decadal running means of annual mean temperature extended back from AD 1950 to 1600 on the basis of sea-ice incidence. (Adapted from Bergthórsson, 1969)

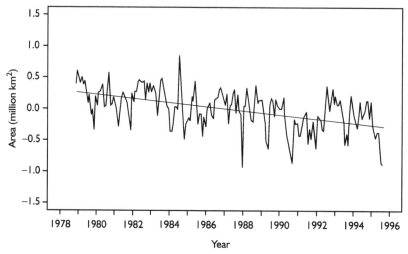

FIGURE 6.8 Monthly anomalies of Arctic ice extent between November 1978 and August 1995. (Adapted from Bjørgo *et al.*, 1997)

years are mentioned for 1405 and 1422–1426. A model describing sea ice/climate/ELA relationships in northern Iceland has been developed (Stötter *et al.*, 1999), providing a calibrated proxy climate record as a basis for palaeoclimatic reconstruction for the Holocene.

Satellite passive microwave sensors are the most appropriate means of investigating the global sea ice cover (e.g. Bjørgo *et al.*, 1997). Global sea-ice cover from 1978 to present is provided by the Nimbus 7 Scanning Multi-channel Microwave Radiometer (SMMR) and the Defence Meteorological Satellite Program (DMSP) Special Sensor Microwave Imager (SSMI). Statistical analysis on the time series (16.8-year observation period) suggests the decreases in Arctic ice extent and ice area to be 4.5 per cent and 5.7 per cent, respectively (Fig. 6.8). From November 1978 to December 1996, the areal extent of sea ice decreased by 2.9 ± 0.4 per cent per decade in the Arctic, with extreme minima recorded in 1990, 1993 and 1995 (Maslanik *et al.*, 1996). In the Antarctic, however, the area of sea ice extent increased by 1.3 ± 0.2 per cent per decade (Cavalieri *et al.*, 1997). The hemispheric asymmetry is consistent with the modelled response of climate warming due to atmospheric increase in CO_2. Situations with minimum ice

cover along the Siberian coast are characterized by warm, windy conditions in May and continued warm weather in June, followed by strong coastal winds in August leading to final break-up and retreat of the pack ice (Serreze *et al.*, 1995).

The National Snow and Ice Data Center (NSIDC) in Boulder, Colorado, USA, currently archives and distributes two sea-ice data-sets produced using the NASA Team Algorithm applied to satellite passive microwave data. These are the DMSPSSM/I Daily and Monthly Polar Gridded Sea Ice Concentrations, and Sea Ice Concentrations from Nimbus-7 SMMR and DMSPSSM/I Passive Microwave data. Both data-sets are derived from 25×25 km gridded brightness temperatures in polar stereographic projection, and provide users with daily- and monthly-averaged sea-ice concentrations for both hemispheres. The NSIDC has also processed daily and monthly total sea-ice extents and total ice-covered areas. These data are provided to investigate interannual variability and trends in sea-ice cover. Total sea-ice area and extent in the northern and southern hemispheres are computed for each day with data. Anomalies are computed from the monthly mean values for November 1978 to November 1987, and February 1988 to December 1996.

Satellite images from the NSIDC World Data Center-A for Glaciology show that between 15 February 1998 and 18 March 1999, the Larsen B ice shelf in Antarctica lost approximately 1839 km².

6.4 Models of future sea-level changes

Future sea-level rise has led to concern because of possible impacts to coastal regions, in the form of loss of land through inundation and erosion, increased frequency of storm floods, and saltwater intrusion. Global mean sea-level is a sensitive indicator of climatic change. Global warming will lead to sea-level rise from the thermal expansion of seawater, the melting of alpine glaciers and polar ice sheets. Several studies have reported rates of annual global sea-level rise during the last century ranging between 0.3 and 3 mm yr^{-1}. The reliability of these data has, however, been questioned due to problems with data quality and spatial and temporal variations in physical processes involved in sea-level variation, such as wind,

ocean currents, river runoff and tectonic movement, of which the latter may introduce a serious source of error. Proxies of palaeo-sea-level changes, geophysical modelling and satellite geodesy may, however, be used to remove isostatic effects. Future sea-level is expected to rise by about 0.5 m by the year 2100, which is 4–7 times faster than present rates. Most global climate models predict a positive net balance in Antarctica because precipitation over the ice sheet may exceed ablation. Because future sea-level projections are quite uncertain, careful monitoring by upgraded tide-gauge networks and satellite geodesy will become important.

According to oxygen isotope ratios in marine sediment cores in relation to astronomical cycles, the present Holocene interglacial period may terminate in a few thousand years (Fig. 6.9). Analyses of solar insolation suggest that the present interglacial may last longer (Berger and Loutre, 1994, 1996), marked by a global fall in temperature and sea-level. Human effects such as the emission of greenhouse gases may, however, lead to the opposite development.

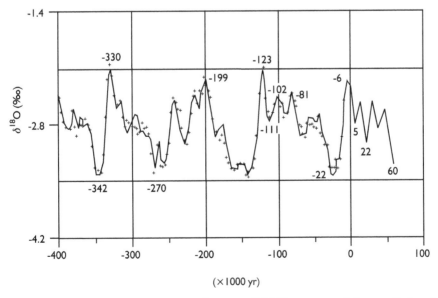

FIGURE 6.9 Long-term climatic variations over the past 400,000 years, and the prediction for the next 60,000 years according to Berger (1988).

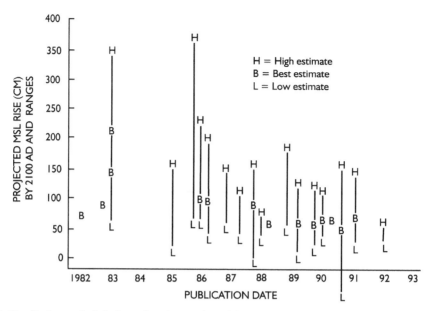

FIGURE 6.10 Predictions of global sea-level rise. (Modified from Pirazzoli, 1996)

In addition to CO$_2$, other so-called green-house gases (methane, ozone, nitrous oxide and chlorofluorocarbons) have shown increasing atmospheric concentrations during the last few decades. Some climate models indicate that the warming produced by the increased greenhouse effect is already taking place and has caused sea-level to rise. There are also predictions that the increase in atmospheric CO$_2$ will continue over the next few decades. This increase is expected to raise the mean global surface air temperature by between 1.5 and 4.5°C. This temperature rise is expected to produce a volumetric expansion of the oceans and melting of continental glaciers, causing global sea-level rise. Observational data related to thermal expansion are too limited to make global estimates. Model simulations, however, suggest an annual rise of the order of 0.3 ± 0.2 mm due to volumetric expansion alone (steric changes). Due to uncertain mass budget measurements and estimates for the Greenland and Antarctic ice sheets, future predictions of global sea-level variations are difficult to make. The Antarctic ice sheet is usually assumed to be in equilibrium or to have a slightly positive mass balance, while

the Greenland ice sheet is probably close to balance (e.g. Pirazzolli, 1996). The majority of the world's alpine glaciers have been retreating over the last century, but their contribution to sea-level rise is limited. It has been roughly estimated that glaciers and ice caps have contributed 0.4 ± 0.3 mm yr^{-1} for the recent sea-level rise (IPCC, 1995).

Estimates of the sea-level rise by the year 2100 vary according to their date of publication (Fig. 6.10). The figure demonstrates a certain uncertainty in the future global sea-level rise. The heavy vertical lines in Fig. 6.10 show when the estimate was made for year 2100, the light vertical lines show the estimates for some year before 2100 but extrapolated to 2100, and the letter B only indicates that the author did not estimate the sea-level range. Since the early 1980s, there has been a decline in the estimated sea-level rise predicted for the next century. According to the Intergovernmental Panel on Climate Change (IPCC, 1995), the 'best estimate' for the year 2100 was a global sea-level rise of 66 cm, with high and low estimates of 110 cm and 30 cm, respectively. In a later estimate, Wigley and Raper (1992) proposed a global sea-level rise of 48 cm (high estimate of 90 cm and low estimate of 15 cm) by the year 2100. The IPCC report of 1995 stated that sea-level is expected to rise as a result of thermal expansion of the world's oceans and melting of alpine glaciers and ice sheets. The 'best estimate' models of climate sensitivity and of ice melt sensitivity to warming, including the effects of future aerosol changes, suggest about a 50 cm increase in global sea-level between 1995 and 2100. This estimate is about 25 per cent lower than the 'best estimate' in the 1995 report due to lower temperature projections and improved climate models. Even if greenhouse gas concentrations were stable by AD 2100, global sea-level is expected to rise at a similar rate for centuries after that time.

Models of future climatic change

7.0 Chapter summary

This chapter offers the reader a synthesis of the theoretical and practical interactions between climate change and glaciers in the past, the potential for an understanding of this inter-action to be applied to modelling of future climate change, and the likely response of global glacier systems to predicted climate change. The chapter evaluates possible future orbitally-induced (Milankovitch) climate change, future climate change and glaciation, and greenhouse gas-induced warming. Furthermore, the chapter evaluates natural versus anthropogenic forcing, energy balance models and glacier variations, and global cir-culation models (GCMs). Finally, future research priorities concerning glaciers and climate change are discussed.

7.1 Orbitally-induced (Milankovitch) climate change

Climate is extremely complex to explain in physical terms. The study of the main climate processes responsible for climate variation requires the construction of mathematical models to describe quantitatively the interact-ing physical elements. Reconstructions of ice-sheet variations over the last million years indi-cate that climate, directly or through feedback mechanisms, is sensitive to variations in the orbital parameters. Climate models have been used to reconstruct the relationship between orbital variations and climate in the past. Few studies have, however, attempted to predict future climate based on variations in the orbital parameters (e.g. Berger and Loutre, 1994). Climate reconstructions for the past 120,000 years and the next tens to hundred thousand years show common features, despite some minor discrepancies. The model reconstruc-tions show, after removing potential anthropo-genic effects, that the world's climate will soon begin to deteriorate towards glacial/stadial conditions (Fig. 7.1).

Oscillatory cooling with progressively colder periods is expected at approximately 5000, 23,000 and 60,000 years from now. The glacial episode around 60,000 years is expected to reach a similar magnitude to the last glacial maximum at 18–20,000 kyr BP. The model simu-lations all indicate that climates as warm as the present Holocene have been and will be rare. A climate such as that of the Holocene will not occur again before 120,000 years from the pre-sent. In the absence of anthropogenic forcing, model result also indicate that the long-term cooling trend which started about 6000 yr BP will continue for the next 5000 years. This first minimum will be followed by stabilization phases approximately 15,000 years and 25,000 years from now. In the northern hemisphere, the next glacial maximum will occur approxi-mately 55,000 yr after present, leading to an average cooling of 0.01 °C per century. This is negligible on a human time-scale, and also within the expected temperature rise for the twenty-first century as a result of increased greenhouse gas concentrations. The build-up of northern hemisphere ice sheets, with an

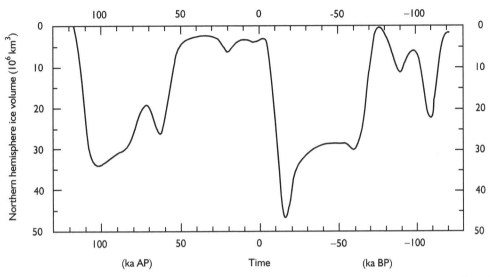

FIGURE 7.1 Predicted ice volume in the northern hemisphere resulting from astronomical and CO_2 forcing. (Modified from Berger and Loutre, 1994)

estimated volume of $27 \times 10^6 \, km^3$ within 55,000 years after present, will correspond to a glacio-eustatic sea-level drop of about 70 m (corresponding to 10 cm per century).

In a model for the next 5000 years, Loutre (1993) assumed that the pre-industrial CO_2 concentration will rise from 280 to 710 ppmv within the next 500 years and then decrease to 450 ppmv and 350 ppmv within 1000 and 1500 years from now, respectively. According to the model, the surface air temperature in the northern hemisphere will increase by 3°C, from 15°C to 18°C, over the next 500 years. The temperature changes are expected to be particularly large north of 65°N during spring and autumn, as a response to a reduction in the extent of ice and snow which is a positive feedback mechanism. Between 500 and 1500 years from now, the surface air temperature will decrease and reach a minimum between 16 and 17°C. According to the model, temperature is expected to increase slightly up to 5000 years from now due to melting of the Greenland ice sheet, which is expected to disappear at the end of the period. Great uncertainties remain, however, between the greenhouse and orbital forcing and future climate change. Three possible scenarios describing the relationship between enhanced greenhouse

warming and orbital forcing have been suggested (Goodess *et al.*, 1992). The first scenario is that a relatively short (ca. 1000 years) period of greenhouse gas-induced warming will be followed by 'natural' glacial–interglacial cycles. The second option is that the next glacial period will be delayed and less severe. The final scenario is that enhanced greenhouse warming will reduce the positive feedback mechanisms to such an extent that future glaciations will be prevented (the irreversible greenhouse effect). In conclusion, modelling results indicate that the pattern and amplitude of climatic conditions in the past million years will be experienced over the next hundreds of thousands of years. However, enhanced greenhouse-gas emissions may cause some deviations from the pattern of orbitally driven climate changes seen in the past.

7.2 Greenhouse gas-induced warming

The IPCC (1992, 1995) developed several scenarios of future greenhouse gas and aerosol emissions based on assumptions of population growth, economic development, land use, technological changes, energy availability and

Box 7.1 Future climate change and glaciation

As the understanding of past global climates and mechanisms of climatic change increases, the need to be able to predict future climate change and the growth and decay of ice sheets becomes increasingly important. Such predictions must, however, be based on geological, climatic and astronomical factors. The history of Late Cenozoic climate change and glaciation demonstrates that the Earth is now experiencing an interglacial phase which began about 10,000 years ago. In Europe, interglacials have been defined when deciduous forests replaced coniferous forests. If Pleistocene interglacials are defined by the presence of deep sea sediment oxygen isotope stage 5e, the interglacial periods lasted in general not longer than 10,000–12,000 years. Based on this evidence, we may be approaching a cooling phase with more glacial conditions. However, if the Little Ice Age temperature depression is repeated in the next centuries, it will compensate in part for anthropogenic warming. Short-term climatic trends give only restricted information. As an example, mean annual global surface temperatures have been rising since the termination of the Little Ice Age, except for a cooling phase from the

1940s to the 1960s. Different proxy data suggest that mean annual temperatures may have dropped 1.5–2°C since the mid-Holocene thermal optimum.

What, then, is the possibility of mountain glaciation occurring in the near future? It has been suggested that initial ice-sheet growth at the start of the last major glaciation (ca. 120,000 yr BP) in high northern latitudes occurred under similar climate conditions as at present (Miller and Vernal, 1992). Optimal ice-growth conditions include a warm northern ocean, strong meridional atmospheric–ocean circulation, and low summer temperatures. Greenhouse warming is suggested to be most significant in the Arctic and during the winter season (Berger, 1978) combined with decreased summer insolation. This may cause snow-line depression and glacier growth in high northern latitudes. Heavy winter precipitation in the late 1980s and early 1990s in western Scandinavia has caused the largest glacier advance recorded on maritime glaciers since the termination of the Little Ice Age (e.g. Nesje *et al.*, 1995). Between 1992 and 1997, Briksdalsbreen, a western outlet glacier from Jostedalsbreen, advanced 322 metres, giving a mean daily advance rate of 18 cm. The largest annual advance at Briksdalsbreen was recorded in 1993/94 (80 m) (Nesje *et al.*, 1995).

fuel mix during the period from 1990 to 2100. The emissions can be used to estimate atmospheric concentrations of greenhouse gases and aerosols and the perturbation of natural radiative forcing. Then coupled atmosphere–ocean climate models can be used to project future climate. Large uncertainties remain, however, and these have been taken into consideration in the range of projections of future mean global temperature change. For the 'best estimate' scenario, including the effects of future increases in aerosols, models project an increase in mean global surface air temperature relative to 1990 of about 2°C by AD 2100. This estimate is about 30 per cent

lower than the 'best estimate' in 1990. The corresponding high and low temperature estimates are 3.5°C and 1.0°C, respectively. The average rate of warming will probably be among the greatest during the last 10,000 years. However, the annual to decadal changes will include significant natural variability, and regional temperature changes may differ substantially from the global mean. All model simulations indicate greater surface warming over the land than in the oceans in winter, with a maximum surface warming in high northern latitudes in winter, little surface warming over the Arctic in the summer, and increased precipitation in high latitudes in winter. Most simulations also show a reduction in the strength of the North Atlantic thermohaline circulation and a reduction in the diurnal temperature range (IPCC, 1995). Higher temperatures will probably cause an intensified hydrological cycle, with more/less severe droughts and floods. Several models suggest an increase in precipitation intensity, with the possibility of more extreme rain-/snowfall events. This could lead to increased winter precipitation on maritime glaciers, causing positive net balance and glacier expansion, such as in western Scandinavia during the 1990s (Nesje et al., 1995).

Ice sheets and glaciated regions play an important role in modulating the global environment. Their growth and retreat change the surface topography and albedo and therefore influence global temperatures and wind patterns. Their behaviour is characterized by feedback mechanisms which may amplify small climatic variations. The almost cyclic behaviour of the ice sheets through time is characteristic of a natural system attempting to establish equilibrium without obtaining stability.

Different views exist about the current state of ice sheets and their possible response to global warming. The Antarctic ice sheet consists of two different parts, which would increase sea-level by 65 m if melted. The East Antarctic ice sheet is continent-based and forms a dome over 4000 m in altitude, drained by a series of ice streams. The West Antarctic ice sheet consists of three domes and rises to about 2000 m. A considerable part of the ice sheet is grounded below present sea-level. The ice sheet is drained by ice streams, several of which flow into ice shelves, at velocities of several hundred metres per year.

It has been suggested that the East Antarctic ice sheet formed some 50–60 million years ago during the Tertiary, and that it became a large ice sheet 38–18 million years ago, after which it has been rather stable. The presence of marine diatoms (the Sirius formation) in a glacial till in the Trans-Antarctic Mountains challenge this view of long-term stability. The diatoms have ages of 2.5–3.0 million years and they have been interpreted to have originally grown in a marine environment in East Antarctica (e.g.

Barrett et al., 1992). This means that the ice sheet must have been much reduced in size at that time and that the ice sheet was more unstable that previously thought. The two models are dramatically different. The former reconstruction suggests that there is little risk of melting the Antarctic ice sheet under possible future global warming. The other, however, implies that the ice sheet may become unstable under future global warming scenarios, and contribute many metres to sea-level rise.

There is some uncertainty relating to the West Antarctic ice sheet, which is grounded below sea-level in several areas. According to Mercer (1978), Hughes (1981) and van der Veen (1987), the ice sheet may be buttressed by the surrounding ice shelves, and a minor sea-level rise or increased melting rates could trigger ice calving, surge and collapse of the ice sheet. Such a collapse, which is known from the Quaternary record, would raise global sea-level by 5–7 m. In West Antarctica, future developments depend on the shape of the subglacial topography. If the calving front retreats into deeper water, the rate of calving will increase, leading to progressive collapse.

The response of the Greenland ice sheet to warming scenarios is also uncertain (e.g. Reeh, 1989). Based on the observed retreat of outlet glaciers in West Greenland over the last century (Weidick, 1984), it is reasonable to expect enhanced melting of the outlet glaciers. Satellite observations show, however, that during the period from 1978 to 1986 the southern part of the ice sheet increased in thickness (Zwally, 1989).

Our understanding of the behaviour of the largest ice masses is poor, because the largest ice sheets are quite inaccessible and so large that accurate calculations of their mass balance are extremely difficult. In addition, the ice sheets are part of a complex system with links to the atmosphere, oceans, biosphere and lithosphere. To be able to predict how they respond to a given forcing, it is necessary to understand the relative importance of the feedback mechanisms to the components in the system. According to Saltzman (1985), we should rather concentrate on using field evidence from terrestrial and adjacent marine areas to constrain the

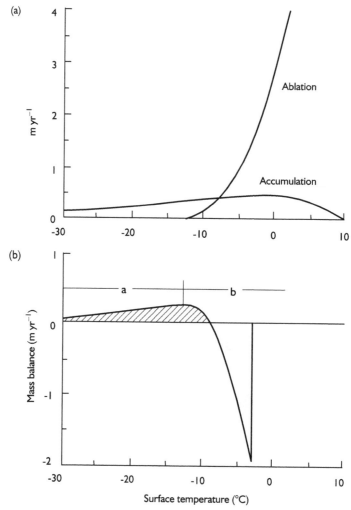

FIGURE 7.2 (a) The relationship of ablation and accumulation to annual surface temperature for an ice sheet. (b) The dependence of net annual mass balance on temperature changes from positive to negative. This complicates the response of ice sheets to climate change. (Adapted from Warrick and Oerlemans, 1990)

models. Therefore field evidence should be used to develop models and theories.

Ice sheets may respond to warming in three main ways: (1) by warming of the ice; (2) by changing the surface mass balance; and (3) by changing sea-level or sea temperature. Ice accumulating on the ice-sheet surface may, however, take several tens of thousands of years to pass through the system. After warming of the air temperature, warmer ice accumulating on the surface will gradually replace colder ice, eventually leading to a lower ice

sheet with a reduced volume. The response time for this adjustment is so long and the effect so small, however, that it is insignificant on a human time-scale. A change in the mass balance is critical to the response of an ice sheet. Commonly for glaciers, more warming leads to more melting. For ice sheets it is not that simple. Figure 7.2 shows how ablation increases from areas where the mean annual temperature is −12°C, to values of over $4\,\mathrm{m\,yr^{-1}}$ where the mean annual temperature is close to 0°C. In cold regions, accumulation is related to the ability of air to hold water vapour. In the interior of ice sheets the mass balance is limited by the snowfall. As a result, in environments colder than approximately $-10°C$, warming will increase the mass balance as precipitation increases, while in milder regions, increased melting will reduce the mass balance. In East Antarctica and northern Greenland, with temperatures mostly below $-10°C$, the mass balance is likely to increase with minor warming. Southern regions of the Greenland ice sheet are likely to be reduced by increased melting. A complicating factor is, however, that melting could lower the marginal areas at the same time as the interior is thickening due to increased snowfall. The result is a steeper surface profile with increased flow velocities, which may counteract the thickening of the interior. In addition, an increase in water depth can trigger iceberg calving, an effect known to produce the highest rates of retreat of marine Quaternary ice sheets (the Laurentide ice sheet over Hudson Bay and fjord glaciers along the coast of Norway). Ice sheets are an important element of the global environmental system, with implications for climate and sea level change. This system is so complex, and has so many feedback loops, that it is extremely difficult to predict the response to future human-induced environmental changes.

The minor mountain glaciers and ice caps contain a small part of the total ice mass on Earth. The retreat of these glaciers subsequent to the Little Ice Age is well documented (e.g. Grove, 1988). The most comprehensive study was made by Meier (1984). He calculated that they contributed 2–4 cm of the sea-level rise during the last century. Since small alpine glaciers respond quickly to climatic change, they are likely to be major contributors to the first phase of possible global warming.

Zuo and Oerlemans (1997) used an ice-flow model to simulate past and future frontal fluctuations of the Pasterze glacier in Austria. Their model experiments suggested that the glacier has been in a non-steady state most of the time, and that the glacier has a response time of 34–50 years. Furthermore, the model shows that the Pasterze glacier will retreat

significantly if warming is taking place, of the order of 2–5 km by the year 2100.

Oerlemans (1997) used historic glacier-length variations to constrain a computer model for future frontal variation of Nigardsbreen, an eastern outlet glacier from Jostedalsbreen in western Norway. When calibrated with past changes, the model predicts an 800 m advance of the glacier front if the mass balance remains as it was between 1962 and 1993. For a uniform heating rate of $0.02\,K\,a^{-1}$, Nigardsbreen will advance slightly until the year 2020, after which a significant retreat will occur, reducing the glacier to 10 per cent of the 1950 volume.

7.3 Natural versus anthropogenic forcing

Since the IPCC (1990, 1992, 1995) scientific assessments, progress has been made towards identifying possible anthropogenic effects on climate. Firstly, model experiments are now incorporating the possible climatic effects of human-induced sulphate aerosols and stratospheric ozone. As a result, the potential climate change signal due to human activities is better defined, although several uncertainties still remain. Secondly, the background climatic variability has been better defined. This is crucial for distinguishing human effects on climate from natural climate variability on decadal to centennial time-scales, including both internal and external components (e.g. changes in solar variability or volcanic dust loading). Thirdly, some progress has been made in the application of pattern-based methods in an attempt to attribute some part of the observed climate changes to human activities (cause–effect relationship). The majority of the studies that have attempted to detect an anthropogenic effect on climate have used mean annual global temperature. Most of these studies indicate that the observed global temperature change during the last century is unlikely only to result from natural temperature fluctuations. However, the records cannot be considered clear evidence of anthropogenic forcing and changes in the Earth's surface temperature.

Recent studies have compared observations with the patterns of temperature change predicted by models as a response to anthropogenic forcing. The background for pattern-based approaches is that different forcing mechanisms show different response patterns. Some studies have compared patterns of temperature variations with model patterns from simulations using changes in CO_2 and anthropogenic sulphate aerosols. The results presented in the IPCC (1995) report indicate that the record of mean global temperature over the past century is probably not entirely of natural origin and that there is a climate response to forcing by greenhouse gases and sulphate aerosols. The ability to quantify the magnitude of this effect is, however, limited as a result of uncertainties in longer-term natural climate variability and the forcing and response patterns to changes in greenhouse gases, aerosols and other human influences (IPCC, 1995 and references therein).

7.4 Energy-balance models and glacier variations

An energy-balance model has been developed by Oerlemans (1988, 1991, 1992). This model calculates the components of the surface energy balance. It takes meteorological data, the area distribution with altitude of ice mass, and parameters defining the global radiation as input values. This model has also been used to calculate mass balance for glaciers in the Austrian Alps (Oerlemans and Hoogendoorn, 1989), southern Norway (Oerlemans, 1992), the Greenland ice sheet (Oerlemans, 1991) and NW Spitzbergen (Fleming *et al.*, 1997). The input data for the model are the annual mass balance, the fraction of meltwater that refreezes instead of running off, the energy balance at the surface, the latent heat of melting, and the rate of precipitation in solid form. The energy exchange between the atmosphere and the glacier surface is found from:

$$B = Q(1 - a) + I_{in} + I_{out} + F_s + F_l \qquad (7.1)$$

where a is the surface albedo, Q is the short-wave radiation reaching the surface, I_{in} and

I_{out} are the incoming and outgoing long-wave radiation, and F_s and F_l are the sensible and latent heat fluxes. The energy budget is divided into several components: solar radiation, long-wave radiation, turbulent energy fluxes and the refreezing of meltwater. The solar radiation reaching the top of the atmosphere is attenuated by absorption and scattering. Cloudiness, solar zenith angle and surface elevation are accounted for. However, the geometry of the glacier surface is not normally accounted for. Surface albedo is dependent on the presence of snow, the distance to the equilibrium line, and the total area of glacier ice exposed during the ablation season. The long-wave component of the energy equation is divided between the outgoing radiation from the glacier surface and the incoming radiation from the atmosphere. The atmospheric long-wave radiation is in two parts: the contribution from a clear sky and that from the clouds. Turbulent fluxes are proportional to the difference between the air and surface temperatures and humidity. There are three approaches by which the required meteorological parameters are made available to the model: (1) the use of long-term climate data; (2) annual fits to measured data; and (3) daily inputs. The annual method has been used most commonly (Oerlemans, 1991, 1992) because daily meteorological data are either lacking or are only available from meteorological stations remote from the glaciers being modelled. This requires the annual temperature to be expressed as a sinusoidal function. Precipitation is defined as constant through the year. Precipitation is assumed to fall as snow when the daily average temperature is below 2°C. All rainfall is assumed to run off. Humidity and cloudiness are constant throughout the year, and set to the mean annual value.

7.5 Global circulation models (GCMs)

To gain a better understanding of atmospheric conditions, we simulate global-scale climatic processes using *global circulation models* (GCMs), which are computer simulations of mathematical models of how the atmosphere operates and interacts with the hydrological cycle (Wright *et al.*, 1993). GCM experiments have provided pertinent information about relationships between atmospheric circulation and mechanisms of climate forcing such as land uplift (Ruddiman and Kutzbach, 1990), vegetation cover (Foley *et al.*, 1994), volcanic activity (Bryson, 1989), desert formation (Joussaume, 1989), ocean circulation (Lautenschlager *et al.*, 1992) and ice cover (Lautenschlager and Herterich, 1990).

GCMs aim to simulate 3D structures and flows of the atmosphere, but even the most powerful computers are unable to model the scale and complexity of the atmospheric circulation system. The surface of the Earth is represented by a grid (the size of the grid cells varies commonly between $4 \times 5°$ and $11.5 \times 11.25°$. In each grid cell, values are plotted representing selected Earth surface parameters. The parameters normally fall into two categories. *Boundary conditions* are surface values of, for example, sea-surface temperatures (SSTs), albedo, radiation, atmospheric transparency, sea-ice cover and topography (e.g. Kutzbach and Ruddiman, 1993). When modelling modern climate, direct measurements of physical parameters comprise the input data, while palaeoclimatic modelling is based on estimates obtained from proxy data. *Dynamic conditions* include flow and can be obtained by parameterization of surface processes (heat and moisture exchange, convection, Coriolis force, shear constants and atmospheric pressure) (Street-Perrott, 1991; Wright *et al.*, 1993).

Since GCMs involve enormous amounts of data and long computer runs by supercomputers, GCMs are only undertaken in a few highly specialist centres such as the National Center for Atmospheric Research (NCAR) in the USA; the Goddard Institute for Space Studies in USA; the UK Meteorological Office; and the Deutsche Klima Rechen Zenter (DKRZ) in Hamburg, Germany.

Several GCM efforts have concentrated on modelling climatic conditions at selected time intervals with appropriate proxy data, for example, the late glacial maximum around 18,000 yr BP. Figure 7.3 shows schematically how input boundary conditions, acting as the basis for GCM experiments, have varied since 18,000 yr BP.

GCM experiments are of two main types: *analogue experiments* and *sensitivity experiments* (Street-Perrott, 1991). Analogue experiments, or so-called 'realistic' experiments, attempt to simulate the prevailing Earth surface conditions as closely as possible for selected time periods. The early CLIMAP and COHMAP models were of this type (CLIMAP Project Members, 1981; COHMAP Members, 1988). Recent models have become more sophisticated, allowing for annual or seasonal variations of input data. Important linkages in the climate system have been demonstrated by the analogue modelling experiments, like the relationship between enhanced summer insolation and monsoon strength in the northern hemisphere, and the deflection of the jet stream by the build-up of the Laurentide ice sheet.

The purpose of the sensitivity experiments, however, is to test the relative importance of the different model components. Input variables are changed individually to elaborate the most responsive factors of the model. By comparing the modelling results with palaeoclimatic reconstructions, it may be possible to discover the most important forcing factor(s). However, some of the modelling outputs have clearly been wrong. Hansen *et al.* (1984) found, for example, that land ice was the most important factor in global cooling during the last glacial maximum, while Broccoli and Manabe (1987) in their experiments found that reduced CO_2 content was of greatest importance.

The most severe limitations with GCMs are the scale problems, such as the topographic generalizations to conform with the large grid cells. Another problem is the varying quality of the input data derived from proxy records. In addition, since GCMs are constructed for fixed time periods, they are considered to be static. Simulation and modelling of highly complex and rapid global and regional climatic changes lies beyond the capacity of existing GCMs.

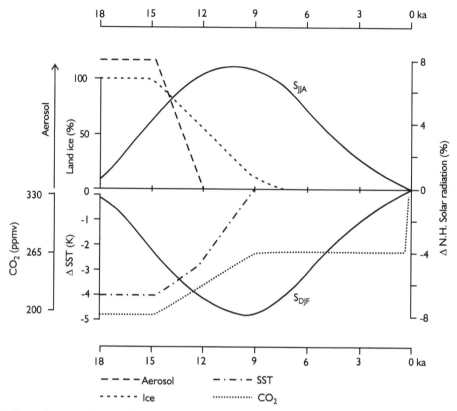

FIGURE 7.3 Boundary conditions for the COHMAP simulation for the last 18,000 years. External forcing is shown for northern hemisphere solar radiation in June–August (S_{JJA}) and December–February (S_{DJF}) as the percentage difference from present-day radiation receipts. Internal boundary conditions include land ice, global mean sea-surface temperatures (SST) expressed as difference from the present, aerosols, and concentrations of CO_2 in ppmv. (Modified from Kutzbach and Webb, 1993)

7.6 Future research priorities

Ice cores are extremely valuable palaeoenvironmental archives, and it seems unlikely that their potential has been exploited fully, as far as both precision and resolution are concerned. The future challenge will be to measure the full range of proxy environmental information stored in ice cores. New ice cores should therefore be obtained in order to validate existing ice-core data. Hopefully, new ice cores from Greenland will penetrate *in situ* Eemian ice.

Differences in magnitude and timing of Late Glacial and Holocene ice advances have the potential to establish climatic gradients and circulation patterns and provide information about underlying climatic factors. Glacial

and climate reconstructions in NW Europe are particularly important in this respect because it is possible to record changes in oceanic and atmospheric circulation patterns in the North Atlantic region. Controls on Late-Glacial deglaciation and the following Holocene period merit further study. So far, there has been a tendency to produce regional correlations of glacier and climate variations. The data compilation presented here, however, demonstrates significant intraregional differences in climate (temperature/precipitation) and glacier response. Future research should therefore focus on establishing representative, multidisciplinary, well-dated records from individual glaciers to study similarities and dissimilarities in glacier response to topographical and climatic parameters (e.g. summer temperature, winter precipitation, wind direction, glacier aspect, radiation). Radiocarbon dates should be given in calendar years in order to study rates of change and rates of sediment accumulation/deposition.

Two particular problems remain to be solved in the context of how various components interact to cause climate change, and how the linkages function between different components in the ocean–atmosphere–terrestrial system. The first is to distinguish between cause and effect, and the second is to establish the precise order of events. Progress will come in both of these important areas as research is further refined (improved dating, better correlations and stratigraphic resolution). Establishment of more multidisiplinary research teams to synthesize proxy data from a range of sources may allow us to come closer to a solution of the complex linkages between the different components of the ocean–atmosphere–terrestrial systems in the past, at present, and in the future.

References

Aber, J.S., Croot, D.G. and Fenton, M.M. (1989) *Glacio-tectonic Landforms and Structures.* Kluwer, Dordrecht.

Adam, S., Pietroniro, A. and Brugman, M.M. (1997) Glacier snow line mapping using ERS-1 SAR imagery. *Remote Sensing of Environment* **61**, 46–54.

Ahlmann, H.W. (1927) Physio-geographical research in the Horung Massif, Jotunheim. 4: Ablation. *Geografiska Annaler* **9**, 35–66.

Alley, R.B. and Anandakrishnan, S. (1995) Variations in melt-layer frequency in the GISP2 ice core: implications for the Holocene summer temperatures in central Greenland. *Annals of Glaciology* **21**, 64–70.

Alley, R.B., Gow, A.J., Meese, D.A., Fitzpatrick, J.J., Waddington, E.D. and Bolzan, J.F. (1997a) Grain-scale processes, folding, and stratigraphic disturbance in the GISP2 ice core. *Journal of Geophysical Research* **102** (C12), 26819–26830.

Alley, R.B. and MacAyeal, D.R. (1994) Ice-rafted debris associated with binge/purge oscillations of the Laurentide Ice Sheet. *Paleoceanography* **9**, 503–511.

Alley, R.B., Meese, D.A., Shuman, C.A., Gow, A.J., Taylor, K.C., Grootes, P.M., White, J.W.C., Ram, M., Waddington, E.D., Mayewski, P.A. and Zelinski, G.A. (1993) Abrupt increase in Greenland snow accumulation at the end of the Younger Dryas event. *Nature* **362**, 527–529.

Alley, R.B., Shumann, C.A., Meese, D.A., Gow, A.J., Taylor, K.C., Cuffey, K.M., Fitzpatrick, J.J., Grootes, P.M., Zielinski, G.A., Ram, M., Spinelli, G. and Elder, B. (1997b) Visual stratigraphic dating of the GISP2 ice core: basis, reproducibility, and application. *Journal of Geophysical Research* **102** (C12), 26367–26381.

Alt, B.T. (1987) Developing synoptic analogs for the extreme mass balance conditions on Queen Elizabeth Island ice caps. *Journal of Climate and Applied Meteorology* **26**, 1605–1623.

Ammann, B. and Lotter, A.F. (1989) Late-Glacial radiocarbon and palynostratigraphy on the Swiss Plateau. *Boreas* **18**, 109–126.

Anda, E., Orheim, O. and Mangerud, J. (1985) Late Holocene glacier variations and climate at Jan Mayen. *Polar Research* **3**, 129–140.

Andersen, B.G. and Mangerud, J. (1989) The last interglacial–glacial cycle in Fennoscandia. *Quaternary International* **3/4**, 21–29.

Andersen, J.L. and Sollid, J.L. (1971) Glacial chronology and glacial geomorphology in the marginal zones of the glaciers Midtdalsbreen and Nigardsbreen, South Norway. *Norsk Geografisk Tidsskrift* **25**, 1–38.

Anderson, R.Y. (1961) Solar–terrestrial climatic patterns in varved sediments. *Annals New York Academy of Science* **95**, 424–439.

Anderson, R.Y. (1992) Possible connection between surface winds, solar activity and the Earth's magnetic field. *Nature* **358**, 51–53.

Andrews, J.T. (1970) *A Geomorphological Study of Postglacial Uplift with Particular Reference to Arctic Canada.* Institute of British Geographers, Special Publication 2.

Andrews, J.T. (1975) *Glacial Systems: An Approach to Glaciers and their Environments.* Duxbury Press, North Scituate, Massachussetts.

Angelis, M.D., Steffensen, J.P., Legrand, M., Clausen, H. and Hammer, C. (1997) Primary aerosol (sea salt and soil dust) deposited in Greenland ice during the last climatic cycle: comparison with east Antarctic records. *Journal of Geophysical Research* **102** (C12), 26681–26698.

Anklin, M., Barnola, J.-M., Schwander, J., Stauffer, B. and Raynod, D. (1995) Processes affecting the CO_2 concentrations measured in Greenland ice. *Tellus* **B47**, 461–470.

Anklin, M., Schwander, J., Stauffer, B., Tschumi, J., Fuchs, A., Barnola, J.M. and Raynaud, D. (1997) CO_2 record betwen 40 and 8 kyr BP from the Greenland Ice Core Project ice core. *Journal of Geophysical Research* **102** (C12), 26539–26545.

Armstrong, T.E., Roberts, B. and Swithinbank, C.W.M. (1973) *Illustrated Glossary of Snow and Ice.* Scott Polar Research Institute, Cambridge.

Astakhov, V.I. (1992) The last glaciation in West Siberia. *Sveriges Geologiska Undersökning, Series Ca* **81**, 21–30.

Astakhov, V.I., Svendsen, J.I., Matioushkov, A., Mangerud, J., Maslenikova, O. and Tveranger, J. (1999) Marginal formations of the last Kara and Barents ice sheets in northern European Russia. *Boreas* **28**, 23–45.

Bahr, D.B. (1995) Simulating iceberg calving with a percolation model. *Journal of Geophysical Research* **100**, 6225–6232.

Baille, M.G.L. (1989) Hekla 3: How big was it? *Endeavor, New Series* **13**, 78–81.

Baldwin, B., Pollack, J.B., Summers, A., Toon, O.B., Sagan, C. and Van Camp, W. (1976) Stratospheric aerosols and climate change. *Nature* **236**, 551.

Ballantyne, C.K. (1989) The Loch Lomond readvance on the Island of Skye, Scotland: glacier reconstruction and palaeoclimatic implications. *Journal of Quaternary Science* 4, 95–108.

Ballantyne, C.K. (1990) The Holocene glacial history of Lyngshalvøya, northern Norway: chronology and climatic implications. *Boreas* 19, 93–117.

Ballantyne, C.K. (1997) Holocene glacier fluctuations in northern Norway. *Palaeoclimate Research* 24, 1–3.

Ballantyne, C.K., McCarroll, D., Nesje, A., Dahl, S.O. and Stone, J.O. (1998) The last ice sheet in north-west Scotland: reconstruction and implications. *Quaternary Science Reviews* 17, 1149–1184.

Bard, E., Hamelin, B. and Fairbanks, R.G. (1990) U–Th ages obtained by mass spectrometry in corals from Barbados: sea level during the past 130,000 years. *Nature* 346, 456–458.

Barlow, L.K., Rogers, J.C., Serreze, M. and Barry, R.G. (1997) Aspects of climate variability in the North Atlantic sector: discussion and relation to the Greenland Ice Sheet Project 2 high-resolution isotope signal. *Journal of Geophysical Research* 102 (C12), 26333–26344.

Barnola, J.-M., Pimienta, P., Raynaud, D. and Korotkevich, Y.S. (1991) CO_2-climate relationship as deduced from the Vostok ice core: a re-examination based on new measurements and on a re-evaluation of the air dating. *Tellus* 43B, 83–90.

Barnola, J.-M., Raynaud, D., Korotkevich, Y.S. and Lorius, C. (1987) Vostok ice core provides 160,000 year record of atmospheric CO_2. *Nature* 329, 408–414.

Barrett, P.J., Adams, C.J., McIntosh, W.C., Swisher, C.C. and Wilson, G.S. (1992) Geochronological evidence supporting Antarctic deglaciation three million years ago. *Nature* 359, 816–818.

Bassinot, F.C., Labeyrie, L.D., Vincent, E., Quidelleur, X., Shackleton, N.J. and Lancelot, Y. (1994) The astronomical theory of climate and the age of the Brunhes–Matuyama magnetic reversal. *Earth and Planetary Science Letters* 126, 91–108.

Baumann, K.-H., Lackschewitz, K.S., Mangerud, J., Spielhagen, R.S., Wolf-Welling, T.C.W., Henrich, R. and Kassens, H. (1995) Reflection of Scandinavian ice sheet fluctuations in Norwegian sea sediments during the the past 150,000 years. *Quaternary Research* 43, 185–197.

Baumgartner, S., Beer, J., Suter, M., Dittrich-Hannen, B., Synal, H.-A., Kubik, P.W., Hammer, C. and Johnsen, S. (1997) Chlorine 36 fallout in the Summit Greenland Ice Core Project ice core. *Journal of Geophysical Research* 102 (C12), 26659–26662.

Becker, B., Kromer, B. and Trimborn, P. (1991) A stable-isotope tree-ring timescale of the Late Glacial/Holocene boundary. *Nature* 353, 647–649.

Beer, J., Johnsen, S.J., Bonani, G., Finkel, R.C., Langway, C.C., Oeschger, H., Stauffer, B., Suter, M. and Woelfli, W. (1992) ^{10}Be as time markers in polar ice cores. In Bard, E. and Broecker, W.S. (eds) *The Last Deglaciation: Absolute and Radiocarbon Chronologies*. NATO ASI Series 1, 2. Springer Verlag, Berlin, 141–153.

Benn, D.I. (1996) Glacier fluctuations in western Scotland. *Quaternary International* 38, 137–147.

Benn, D.I. and Evans, D.J.A. (1998) *Glaciers and Glaciation*. Arnold, London.

Benson, L.V., May, H.M., Antweiler, R.C., Brinton, T.I., Kashgarin, M., Smoot, J.P. and Lund, S.P. (1998) Continuous lake-sediment records of glaciation in the Sierra Nevada between 52,600 and 12,500 ^{14}C yr BP. *Quaternary Research* 50, 113–127.

Berger, A. (1978) Long-term variations of calorific insolation resulting from the Earth's orbital elements. *Quaternary Research* 9, 139–167.

Berger, A. (1988) Milankovitch theory and climate. *Reviews of Geophysics* 26, 624–657.

Berger, A. and Loutre, M.F. (1994) The climate of the next 100,000 years. Invited paper, *Colloque GEOPROSPECTIVE*, Paris, April 18–19, 1–9.

Berger, A. and Loutre, M.F. (1996) Modelling the climate response to astronomical and CO_2 forcings. *Comptes Rendus de l'Académie des Sciences Serie 2 Fascicule A*, 1–16.

Bergthórsson, P. (1969) An estimate of drift ice and temperature in Iceland in 1000 years. *Jökull* 19, 94–101.

Beschel, R.E. (1950) Flechten als Altersmasstab rezenter Moränen. *Zeitschrift für Gletscherkunde und Glazialgeologie* 1, 152–161.

Beschel, R.E. (1957) Lichenometrie im Gletschervorfeld. *Jahrbuch, Verein zum Schutz der Alpenplanzen und-tierre (München)* 22, 164–185.

Beshel, R.E. (1961) Dating rock surfaces by lichen growth and its application to glaciology and physiography (lichenometry). In Raasch, G.O. (ed.) *Geology in the Arctic* 2. University of Toronto Press, Toronto, 1044–1062.

Bickerton, R.W. and Matthews, J.A. (1993) 'Little Ice Age' variations of outlet glaciers from the Jostedalsbreen ice-cap, southern Norway: a regional lichenometric-dating study of ice-marginal moraine sequences and their climatic significance. *Journal of Quaternary Science* 8, 45–66.

Biscaye, P.E., Grousset, F.E., Revel, M., Van der Gaast, S., Zielinski, G.A., Vaars, A. and Kukla, G. (1997) Asian provenance of glacial dust (stage 2) in the Greenland Ice Sheet Project 2 Ice Core, Summit, Greenland. *Journal of Geophysical Research* 102 (C12), 26765–26781.

Björck, S., Ingolfsson, O., Haflidason, H., Hallsdottir, M. and Anderson, N.J. (1992) Lake Torfadalsvatn: a high resolution record of the North Atlantic ash zone I and the last glacial–interglacial environmental changes in Iceland. *Boreas* 21, 15–22.

Björck, S., Walker, M.J.C., Cwynar, L.C., Johnsen, S., Knudsen, K.-L., Lowe, J.J., Wolfarth, B. and INTIMATE Members (1998) An event stratigraphy for the last termination in the North Atlantic region based on the Greenland ice-core record: a proposal by the INTIMATE group. *Journal of Quaternary Science* 13, 283–292.

Bjørgo, E., Johannessen, O.M. and Miles, M.W. (1997) Analaysis of merged SMMMR- SSMI time series of Arctic and Antarctic sea ice parameters 1978–1995. *Geophysical Research Letters* 24, 413–416.

Blunier, T., Chappelaz, J., Schwander, J., Stauffer, B. and Raynaud, D. (1995) Variations in atmospheric methane concentration during the Holocene epoch. *Nature* 374, 46–49.

Bond, G.C., Broecker, W., Johnsen, S., Mcmanus, J., Labeyrie, L., Jouzel, J. and Bonani, G. (1993) Correlations between climate records from the North Atlantic sediments and Greenland ice. *Nature* 365, 143–147.

Bond, G.C. and Lotti, R. (1995) Iceberg discharges into the North Atlantic on millennial timescales during the last glaciation. *Science* 267, 1005–1010.

Bondarev, L.G., Gobedzhishvili, R.G. and Solomina, O.N. (1997) Fluctuations of local glaciers in the southern ranges of the former USSR: 18,000–8000 BP. *Quaternary International* **38/39**, 103–108.

Bondevik, S., Mangerud, J., Ronnert, L. and Salvigsen, O. (1995) Postglacial sea-level history of Edgeøya and Baerentsøya, eastern Svalbard. *Polar Research* **14**, 153–180.

Boulton, G.S. (1975) Processes and patterns of subglacial sedimentation: a theoretical approach. In Wright, A.E. and Moseley, F. (eds) *Ice Ages: Ancient and Modern*. Seel House Press, Liverpool, 7–42.

Boulton, G.S. (1979) Processes of glacier erosion on different substrata. *Journal of Glaciology* **23**, 15–38.

Boulton, G.S. (1986) Push moraines and glacier contact fans in marine and terrestrial environments. *Sedimentology* **33**, 677–698.

Boulton, G.S. and Eyles, N. (1979) Sedimentation by valley glaciers, a model and genetic classification. In Schluchter, C. (ed.) *Moraines and Varves*. Balkema, Rotterdam, 11–23.

Boulton, G.S. and Jones, A.S. (1979) Stability of temperate ice caps and ice sheets resting on beds of deformable sediment. *Journal of Glaciology* **24**, 29–43.

Boulton, G.S., Mason, P., Ballantyne, C.K., Karlén, W., Matthews, J.A. and Nesje, A. (1997) Holocene glacier fluctuations in Scandinavia. *Palaeoclimate Research* **24**, 5–33.

Boulton, G.S., Smith, G.O., Jones, A.S. and Newson, J. (1985) Glacial geology and glaciology of the last mid-latitude ice sheets. *Journal of the Geological Society of London* **142**, 447–474.

Boyle, E.A. (1988) Cadmium chemical tracer of deepwater. *Palaeoceanography* **3**, 471–489.

Bradley, R.S. (1978) Volcanic dust influence on glacier mass balance at high latitudes. *Nature* **271**, 736–738.

Bradley, R.S. (1985) *Quaternary Paleoclimatology*. Allen & Unwin, London.

Bradley, R.S. (1988) The explosive volcanic eruption signal in Northern Hemisphere continental temperature records. *Climate Change* **12**, 221–243.

Bradley, R.S. (1990) Holocene paleoclimatology of the Queen Elizabeth Islands, Canadian high Arctic. *Quaternary Science Reviews* **9**, 365–384.

Bradley, R.S. and Eddy, J. (1991) Records of past global changes. In Bradley, R.S. (ed.) *Global Changes of the Past*. UCAR/Office for Interdisiplinary Earth Studies, Boulder, Colorado, 5–9.

Bradley, R.S. and Jones, P.D. (1993) 'Little Ice Age' summer temperature variations: their nature and relevance to recent global warming trends. *Holocene* **3**, 367–376.

Braithwaite, R.J. (1995) Positive degree-day factors for ablation on the Greenland ice sheet studied by energy-balance modeling. *Journal of Glaciology* **41**, 153–160.

Brazier, V., Owens, I.F., Soons, J.M. and Sturman, A.P. (1992) Report on the Franz Josef Glacier. *Zeitschrift für Geomorphologie Supp.* **86**, 35–49.

Briffa, K.R., Jones, P.D., Schweingruber, F.H. and Osborn, T.J. (1998) Influence of volcanic eruptions on Northern Hemisphere summer temperature over the past 600 years. *Nature* **393**, 450–455.

Broadbent, N.D. (1979) *Coastal Resources and Settlement Stability. A Critical Analysis of a Mesolithic Site Complex in Northern Sweden. Aun 3*. Archaeological Studies Institute of Northern European Archaeology, University of Uppsala, Uppsala Borgströms Tryckeri.

Broccoli, A.J. and Manabe, S. (1987) The influence of continental ice, atmospheric CO_2 and land albedo on the climate of the last glacial maximum. *Climate Dynamics* **1**, 87–89.

Broecker, W.S. (1992) Defining the boundary of the Late-Glacial isotope episodes. *Quaternary Research* **38**, 135–138.

Broecker, W.S., Bond, G. and Klas, M. (1990) A salt oscillation in the glacial North Atlantic? 1. The concept. *Paleoceanography* **5**, 469–477.

Broecker, W.S. and Denton, G.H. (1990) The role of ocean-atmosphere reorganisations in glacial cycles. *Quaternary Science Reviews* **9**, 305–341.

Broecker, W.S., Kennett, J.P., Flower, B.P., Teller, J.T., Trumboe, S., Bonani, G. and Wölfli, W. (1989) Routing of meltwater from the Laurentide Ice Sheet during the Younger Dryas cold episode. *Nature* **341**, 318–321.

Broecker, W.S., Peteet, D.M. and Rind, D. (1985) Does the ocean–atmosphere system have more than one stable mode of operation? *Nature* **315**, 21–25.

Brunnberg, L. (1995) Clay-varve chronology and deglaciation during the Younger Dryas and Preboreal in the easternmost part of the Middle Swedish Ice Marginal Zone. *Quaternaria Serie A*, **2**, 1–94.

Bryson, R.A. (1989) Late Quaternary volcanic modulation of Milankovitch climate forcing. *Theoretical Applied Climatology* **39**, 115–125.

Calkin, P.E. (1988) Holocene glaciation of Alaska (and adjoining Yukon territory, Canada). *Quaternary Science Reviews* **7**, 159–184.

Calkin, P.E., Kaufmann, D.S., Przybyl, B.,Whitford, W.B. and Peck, B.J. (1998) Glacier regimes, periglacial landforms, and Holocene climate change in the Kigluaik Mountains, Seward Peninsula, Alaska, USA *Arctic and Alpine Research* **30**, 154–165.

Casassa, G., Brecher, H., Rivera, A. and Aniya, M. (1997) A century-long recession record of Glaciar O'Higgins, Chilean Patagonia. *Annals of Glaciology* 24, 106–110.

Cavalieri, D.J., Gloersen, P., Parkinson, C.L., Comiso, J.C. and Zwally, H.J. (1997) Observed hemispheric asymmetry in global sea ice changes. *Science* **278**, 1104–1106.

Chappelaz, J., Barnola, J.M., Raynod, D., Korotkevitch, Y.S. and Lorius, C. (1990) Ice-core record of atmosphere methane over the past 160,000 years. *Nature* **345**, 127–131.

Chappelaz, J., Brook, E., Blunier, T. and Malaizé, B. (1997) CH_4 and $\delta^{18}O$ of O_2 records from Antarctic and Greenland ice: a clue for stratigraphic disturbance in the bottom part of the Greenland Ice Core Project and the Greenland Ice Sheet Project 2 ice cores. *Journal of Geophysical Research* **102** (C12), 26547–26557.

Charleson, R., Lovelock, J., Andreae, M. and Warren, S. (1987) Oceanic phytoplankton, atmospheric sulphur, cloud albedo and climate. *Nature* **326**, 655–661.

Chen, J. and Funk, M. (1990) Mass balance of Rhonegletscher during 1881/83–1986/87. *Journal of Glaciology* **36**, 199–209.

Clapperton, C.M. (1993) Glacier advances in the Andes at 12,500–10,000 yr BP: implications for mechanisms of Late-glacial climate change. *Journal of Quaternary Science* **8**, 197–215.

Clapperton, C.M. (1995) Fluctuations of local glaciers at the termination of the Pleistocene: 18–8 ka BP. *Quaternary International* **28**, 41–50.

Clapperton, C.M. (1997) Fluctuations of local glaciers 30–8 ka BP: overview. *Quaternary International* **38/39**, 3–6.

Clapperton, C.M. (1998) Late Quaternary glacier fluctuations in the Andes: testing the synchrony of global change. *Quaternary Proceedings* **6**, 65–73.

Clapperton, C.M., Clayton, J.D., Benn, D.I., Marden, C.J. and Argollo, J. (1997) Late Quaternary glacier advances and palaeolake highstands in the Bolivian Altiplano. *Quaternary International* **38/39**, 49–59.

Clapperton, C.M. and Sugden, D.E. (1988) Holocene glacier fluctuations in South America and Antarctica. *Quaternary Science Reviews* **7**, 185–198.

Clapperton, C.M., Sugden, D.E. and Pelto, M. (1989) Relationship of land terminating and fjord glaciers to Holocene climatic change, South Georgia, Antarctica. In Oerlemans, J. (ed.) *Glacier Fluctuations and Climatic Change.* Kluwer Academic Publishers, Dordrecht, 57–75.

Clark, D.H. and Gillespie, A.R. (1997) Timing and significance of Late-Glacial and Holocene cirque glaciation in the Sierra Nevada, California. *Quaternary International* **38/39**, 21–38.

Clark, J.A. (1980) A numerical model of worldwide sea level changes on a viscoelastic earth. In Mörner, N.A. (ed.) *Earth Rheology, Isostasy and Eustasy.* Wiley, Chichester, 525–534.

Clark, J.A., Farrell, W.E. and Peltier, W.R. (1978) Global changes in postglacial sea level, a numerical calculation. *Quaternary Research* **9**, 265–287.

Clark, P.U., Clague, J.J., Curry, B.B., *et al.* (1993) Initiation and development of the Laurentide and Cordilleran ice sheets following the last interglaciation. *Quaternary Science Reviews* **12**, 79–114.

Clark, P.U., Licciardi, J.M., MacAyeal, D.R. and Jenson, J.W. (1996) Numerical reconstruction of a soft-bedded Laurentide ice sheet during the last glacial maximum. *Geology* **24**, 679–682.

CLIMAP Project Members (1981) *Seasonal Reconstructions of the Earth's Surface at the Last Glacial Maximum.* Geological Society of American Map and Chart Series, MC-36.

Cogley, J.G. and Adams, W.P. (1998) Mass balance of glaciers other than ice sheets. *Journal of Glaciology* **44**, 315–325.

COHMAP Members (1988) Climatic changes of the last 18,000 years: observations and model simulations. *Science* **241**, 1043–1052.

Colman, S.M. (1981) Rock-weathering rates as a function of time. *Quaternary Research* **15**, 250–264.

Cromack, M. (1991) Interpretation of laminated sediments from glacier-fed lakes, northwest Spitsbergen. *Norsk Geologisk Tidsskrift* **71**, 129–132.

Crowell, J.C. (1978) Gondwanan glaciation, cyclothems, continental positioning and climate change. *American Journal of Science* **278**, 1345–1372.

Cuffey, K.M., Alley, R.B., Grootes, P.M. and Anandakrishnan, S. (1992) Towards using borehole temperatures to calibrate an isotopic paleothermometer in central Greenland. *Palaeogeography, Palaeoclimatology, Palaeoecology* **98**, 265–268.

Cuffey, K.M., Alley, R.B., Grootes, P.M., Bolzan, J.F. and Anandakrishnan, S. (1994) Calibration of the delta ^{18}O isotopic paleothermometer for central Greenland, using borehole temperatures. *Journal of Glaciology* **40**, 341–349.

Cuffey, K.M. and Clow, G.D. (1997) Temperature, accumulation, and ice sheet elevation in central Greenland through the last deglacial transition. *Journal of Geophysical Research* **102** (C12), 26383–26396.

Cuffey, K.M., Clow, G.D., Alley, R.B., Stuiver, M., Waddington, E.D. and Saltus, R.W. (1995) Large Arctic temperature change at the Wisconsin–Holocene glacial transition. *Science* **270**, 455–458.

Dahl-Jensen, D., Mosegaard, K., Gundestrup, N., Clow, G.D., Johnsen, S.J., Hansen, A.W. and Balling, N. (1998) Past temperatures directly from the Greenland ice sheet. *Science* **282**, 268–271.

Dahl, S.O. and Nesje, A. (1992) Paleoclimatic implications based on equilibrium-line altitude depressions of reconstructed Younger Dryas and Holocene cirque glaciers in inner Nordfjord, western Norway. *Palaeogeography, Palaeoclimatology, Palaeoecology* **94**, 87–97.

Dahl, S.O. and Nesje, A. (1994) Holocene glacier fluctuations at Hardangerjøkulen, central southern Norway; a high-resolution composite chronology from lacustrine and terrestrial deposits. *Holocene* **4**, 269–277.

Dahl, S.O. and Nesje, A. (1996) A new approach to calculating Holocene winter precipitation by combining glacier equilibrium-line altitudes and pine-tree limits: a case study from Hardangerjøkulen, central southern Norway. *Holocene* **6**, 381–398.

Dahl, S.O., Nesje, A. and Øustedal, J. (1997) Cirque glaciers as morphological evidence for a thin Younger Dryas ice sheet in east-central southern Norway. *Boreas* **26**, 161–180.

Dansgaard, W., Johnsen, S.J., Reeh, N., Gundestrup, N., Clausen, H.B. and Hammer, C.U. (1975) Climatic change, Norsemen and modern man. *Nature* **255**, 24–28.

Dansgaard, W., Johnsen, S.J., Clausen, H.B., Dahl-Jensen, D., Gundestrup, N., Hammer, C.U. and Oeschger, H. (1984) North Atlantic oscillations revealed by deep Greenland ice cores. *Geophysical Monographs* **29**, 288–298.

Dansgaard, W., Johnsen, S.J., Clausen, H.B., Dahl-Jensen, D., Gundestrup, N.S., Hammer, C.U., Hvidberg, C.S., Steffensen, J.P., Sveinbjörnsdottir, A.E., Jouzel, J. and Bond, G. (1993) Evidence for general instability of past climate from a 250-kyr ice-core record. *Nature* **364**, 218–220.

Dansgaard, W. and Oeschger, H. (1989) Past environmental long-term records from the Arctic. In Oeschger, H. and Langway, C.C. Jr (eds) *The Environmental Record in Glaciers and Ice Sheets.* John Wiley, Chichester, 287–317.

Dansgaard, W., White, J.W.C. and Johnsen, S.J. (1989) The abrupt termination of the Younger Dryas climatic event. *Nature* **339**, 532–534.

Deevey, E.S. and Flint, R.F. (1957) Postglacial hypsithermal interval. *Science* **125**, 182–184.

De Geer, G. (1912) A geochronology of the last 12,000 years. *Congrès de Géologie International* **11**, 241–253.

Denton, G.H., Armstrong, R.L. and Stuiver, M. (1971) The Late Cenozoic glacial history of Antarctica. In Turekian, K. (ed.) *The Late Cenozoic Glacial Ages.* Yale University Press, New Haven, CT, 267–306.

Denton, G.H. and Hughes, T.J. (1981) *The Last Great Ice Sheets.* John Wiley and Sons, New York.

Denton, G.H. and Karlén, W. (1973) Holocene climatic variations: their pattern and possible causes. *Quaternary Research* **3**, 155–205.

Denton, G.H., Sugden, D.E., Marchant, D.R., Hall, B.L. and Wich, T.I. (1993) East Antarctic ice sheet sensitivity to Pliocene climatic change from a Dry Valley perspective. *Geografiska Annaler* **75A**, 155–204.

Derbyshire, E. and Owen, L.A. (1997) Quaternary glacial history of the Karakoram Mountains and northwest Himalayas: a review. *Quaternary International* **38/39**, 85–102.

Deslodges, J.R. (1994) Varve deposition and the sediment yield record at the small lakes of the southern Canadian Cordillera. *Arctic and Alpine Research* **26**, 130–140.

Dowdeswell, J.A., Hagen, J.O., Björnsson, H., Glazovsky, A.F., Harrison, W.D., Holmlund, P., Jania, J., Koerner, R.M., Lefauconnier, B., Ommanney, C.S.L. and Thomas, R.H. (1997) The mass balance of circum-Arctic glaciers and recent climate change. *Quaternary Research* **48**, 1–14.

Dowdeswell, J.A., Hamilton, G.S. and Hagen, J.O. (1991) The duration of the active phase on surge-type glaciers: contrasts between Svalbard and other regions. *Journal of Glaciology* **37**, 388–400.

Drewry, D.J. (1978) Aspects of the early evolution of West Antarctic ice. In van Zinderen Bakker, E.M. (ed.) *Antarctic Glacial History and World Palaeoenvironments.* Balkema, Rotterdam, 25–32.

Dugmore, A.J. (1989) Tephrachronological studies of Holocene glacier fluctuations in south Iceland. In Oerlemans, J. (ed.) *Glacier Fluctuations and Climatic Change.* Kluwer Academic Publishers, Dordrecht, 37–55.

Dyke, A.S., Andrews, J.T. and Miller, G.H. (1982) *Quaternary Geology of Cumberland Peninsula, Baffin Island, District of Franklin.* Geological Survey of Canada Memoir 403.

Echelmeyer, K. and Harrison, W.D. (1990) Jacobshavn Isbræ, west Greenland, seasonal variations in velocity – or lack thereof. *Journal of Glaciology* **36**, 82–88.

Echelmeyer, K. and Wang, Z. (1987) Direct observation of basal sliding and deformation of basal drift at subfreezing temperatures. *Journal of Glaciology* **33**, 83–98.

Echelmeyer, K. and Zhong Xiang, W. (1987) Direct observation of basal sliding and deformation of basal drift at sub-freezing temperatures. *Journal of Glaciology* **33**, 83–98.

Elven, R. (1978) Subglacial plant remains from the Omnsbreen glacier area, southern Norway. *Boreas* **7**, 83–89.

Elverhøi, A., Dowdeswell, J.A., Funder, S., Mangerud, J. and Stein, R. (1998) Glacial and oceanic history of the Polar North Atlantic margins: an overview. *Quaternary Science Reviews* **17**, 1–10.

England, J. (1986) A paleoglaciation level for north-central Ellesmere Island, NWT Canada. *Arctic and Alpine Research* **18**, 217–222.

England, J. (1998) Support for the Innuitan ice sheet in the Canadian High Arctic during the last glacial maximum. *Journal of Quaternary Science* **13**, 275–280.

Erikstad, L. and Sollid, J.L. (1986) Neoglaciation in South Norway using lichenometric methods. *Norsk Geografisk Tidsskrift* **40**, 85–105.

Etheridge, D.M., Steele, L.P., Langenfelds, R.L., Francey, R.J., Barnola, J.M. and Morgan, V.I. (1996) Natural and anthropogenic changes in atmospheric CO_2 over the last 1000 years from air in Antarctic ice and firn. *Journal of Geophysical Research* **101**, 4115–4128.

Evans, D.J.A., Butcher, C. and Kirthisingha, A.V. (1994) Neoglaciation and early 'Little Ice Age' in western Norway, lichenometric evidence from the Sandane area. *Holocene* **4**, 278–289.

Evans, D.J.A. and England, J. (1992) Geomorphological evidence of Holocene climatic change from northwest Ellesmere Island, Canadian high Arctic. *Holocene* **2**, 148–158.

Eyles, C.H., Eyles, N. and Lagoe, M.B. (1991) The Yakataga Formation, a late Miocene to Pleistocene record of temperate glacial marine sedimentation in the Gulf of Alaska. In Anderson, J.B. and Ashley, G.M. (eds) *Glacial-Marine Sedimentation, Palaeoclimatic Significance.* Geological Society of America Special Paper 261, 159–180.

Fairbanks, R.G. (1989) A 17,000-year glacio-eustatic sea-level record: influence of glacial melting rates on the Younger Dryas event and deep-ocean circulation. *Nature* **342**, 637–642.

Finkel, R.C. and Nishiizumi, K. (1997) Berryllium-10 concentrations in the Greenland Ice Sheet Projects 2 ice core from 3–40 ka. *Journal of Geophysical Research* **102** (C12), 26699–27706.

Fischer, H., Wahlen, M., Smith, J., Mastroianni, D. and Deck, B. (1999) Ice core records of atmospheric CO_2 around the last three glacial terminations. *Science* **283**, 1712–1714.

Fisher, D.A., Reeh, N. and Langley, K. (1985) Objective reconstructions of the Late Wisconsin Laurentide ice sheet and the significance of deformable beds. *Geographie Physique et Quaternaire* **39**, 229–238.

Fitzsimons, S.J. (1997) Late-Glacial and early Holocene glacier activity in the southern Alps, New Zealand. *Quaternary International* **38/39**, 69–76.

Fleming, K.M., Dowdeswell, J.A. and Oerlemans, J. (1997) Modelling the mass balance of nortwest Spitsbergen glaciers and response to climatic change. *Annals of Glaciology* **24**, 203–210.

Flint, R.F. (1971) *Glacial and Quaternary Geology.* Wiley, New York.

Foley, J.A., Kutzbach, J.E., Coe, M.T. and Levis, S. (1994) Feedbacks between climate and boreal forest during the Holocene epoch. *Nature* **371**, 52–54.

Forman, S.L. (1989) Late Weichselian glaciation and deglaciation of Forlandssundet area, western Spitsbergen, Svalbard. *Boreas* **18**, 51–60.

Forman, S.L., Lubinski, D.J., Zeeberg, J.J., Polyak, L., Miller, G.H., Matishov, G. and Tarasov, G. (1999) Postglacial emergence and Late Quaternary glaciation on northern Novaya Zemlja, Arctic Russia. *Boreas* **28**, 133–145.

Forsström, L. and Punkari, M. (1997) Initiation of the last glaciation in northern Europe. *Quaternary Science Reviews* **16**, 1197–1215.

Francou, B., Ribstein, P., Saravia, R. and Tiriau, E. (1995) Monthly balance and water discharge on an intertropical glacier: the Zongo Glacier, Cordilleran Real, Bolvia, 16°S. *Journal of Glaciology* **41**, 61–67.

Fuhrer, K., Wolff, E. and Johnsen, S. (1998) Timescales for dust variability in the GRIP ice core during the last 100 000 years. *PAGES Open Science Meeting. Poster Abstracts,* 65.

Funk, M., Morelli, R. and Stahel, W. (1997) Mass balance of Griesgletscher 1961–1994: different methods of determination. *Zeitschrift für Gletscherkunde und Glazialgeologie* **33**, 41–56.

Furbish, D.J. and Andrews, J.T. (1984) The use of hypsometry to indicate long-term stability and response of valley glaciers to change in mass transfer. *Journal of Glaciology* **30**, 199–211.

Gellatly, A.F., Chinn, T.J.H. and Röthlisberger, F. (1988) Holocene glacier variations in New Zealand: a review. *Quaternary Science Reviews* **7**, 227–242.

Gillespie, A. and Molnar, P. (1995) Asynchronous maximum advance of mountain and continental glaciers. *Reviews of Geophysics* **33**, 311–364.

Gilliland, R.L. (1982) Solar, volcanic, and CO_2 forcing of recent climate change. *Climatic Change* **4**, 111–131.

Gilliland, R.L. (1989) Solar evolution. *Palaeogeography, Palaeoclimatology, Palaeoecology* **75**, 35–55.

Giraudi, C. and Frezzotti, M. (1997) Late Pleistocene glacial events in the central Apennines, Italy. *Quaternary Research* **48**, 280–290.

Glen, J.W. (1955) The creep of polycrystalline ice. *Proceedings of the Royal Society, Series A* **228**, 519–538.

Goodess, C.M., Palutikof, J.P. and Davies, T.D. (eds) (1992) *The Nature and Causes of Climate Change*. Studies in Climatology Series, Lewis Publishers, Belhaven Press, London.

Gordon, J.E. (1980) Recent climatic trends and local glacier margin fluctuations in West Greenland. *Nature* **284**, 157–159.

Gray, J.M. and Coxon, P. (1991) The Loch Lomond Stadial glaciation in Britain and Ireland. In Ehlers, J., Gibbard, P.L. and Rose, J. (eds) *Glacial Deposits in Great Britain and Ireland*.Balkema, Rotterdam, 89–105.

Greuell, W. (1992) Hintereisferner, Austria: mass balance reconstruction and numerical modelling of historical length variation. *Journal of Glaciology* **38**, 233–244.

Griffey, N.J. and Worsley, P. (1978) The pattern of Neoglacial glacier variations in the Okstindan region of Northern Norway during the last three millennia. *Boreas* **7**, 1–17.

Grootes, P.M. and Stuiver, M. (1997) Oxygen 18/16 variability in Greenland snow and ice with 10^{-3}–10^5 year time resolution. *Journal of Geophysical Research* **102** (C12), 26455–26470.

Grosswald, M.G. (1980) Late Weichselian ice sheet of northern Eurasia. *Quaternary Research* **13**, 1–32.

Grosswald, M.G. (1993) Extent and melting history of the late Weichselian ice sheet, the Barents-Kara continental margin. In Peltier, R.W. (ed.) *Ice in the Climate System*. NATO ASI Series I: Global Environmental Change. Springer Verlag, Berlin, 1–20.

Grosswald, M.G. (1998) Late-Weichselian ice sheets in Arctic and Pacific Siberia. *Quaternary International* **45/46**, 3–18.

Grosswald, M.G. and Hughes, T.J. (1995) Paleoglaciology's grand unsolved problem. *Journal of Glaciology* **41**, 313–332.

Grove, J.M. (1972) The incidence of landslides, avalanches, and floods in western Norway during the Little Ice Age. *Arctic and Alpine Research* **4**, 131–138.

Grove, J.M. (1988) *The Little Ice Age*. Methuen, London.

Grove, J.M. (1997) The spatial and temporal variations of glaciers during the Holocene in the Alps, Pyrenees, Tatra and Caucasus. In Frenzel, B., Boulton, G.S., Gläser, B. and Huckriede, U. (eds) Glacier fluctuations during the Holocene. *Palaeoclimate Research* **24**, 95–103.

Grove, J.M. and Battagel, A. (1983) Tax records from western Norway, as an index of Little Ice Age environmental and economic deterioration. *Climatic Change* **5**, 265–282.

Grove, J.M. and Gellatly, A.F. (1997) Glacier fluctuations in the Pyrenees in the Little Ice Age and mid-Holocene. In Frenzel, B., Boulton, G.S., Gläser, B. and Huckriede, U. (eds) *Glacier Fluctuations During the Holocene*. *Palaeoclimate Research* **24**, 67–83.

Gudmundsson, H.J. (1997) A review of the Holocene environmental history of Iceland. *Quaternary Science Reviews* **16**, 81–92.

Gulliksen, S., Birks, H.H., Possnert, G. and Mangerud, J. (1998) A calendar age estimate of the Younger Dryas–Holocene boundary at Kråkenes, western Norway. *Holocene* **8**, 249–259.

Gupta, S.K., Sharma, P. and Shan, S.K. (1992) Constraints on the ice sheet thickness over the Tibet during the last 40,000 years. *Journal of Quaternary Science* **7**, 283–290.

Haakensen, N. (1989) Akkumulasjon på breene i Sør-Norge vinteren 1988–89. *Været* **13**, 91- 94.

Haas, J.N., Richoz, I., Tinner, W. and Wick, L. (1998) Synchronous Holocene climatic oscillations recorded on the Swiss Plateau and at timberline in the Alps. *Holocene* **8**, 301–309.

Haeberli, W. and Beniston, M. (1998) Climate change and its impacts on glaciers and premafrost in the Alps. *Ambio* **27**, 258–265.

Haeberli, W., Müller, P, Alean, P. and Bösch, H. (1989) Glacier changes following the Little Ice Age – a survey of the international data basis and its perspectives. In Oerlemans, J. (ed.) *Glacier Fluctuations and Climatic Change*. Kluwer Academic Publishers, Dordrecht, 77–101.

Hagen, J.O. and Liestøl, O. (1990) Long term glacier mass balance investigations in Svalbard 1950-1988. *Annals of Glaciology* **14**, 102–106.

Hallberg, G.R. (1986) Pre-Wisconsin glacial stratigraphy of the central plains region in Iowa, Nebraska, Kansas, and Missouri. In Sibrava, V., Bowen, D.Q. and Richmond, G.M. (eds) *Quaternary Glaciation in the Northern Hemisphere. Quaternary Science Reviews* **5**, 11–15.

Hallet, B. (1981) Glacier abrasion and sliding: their dependence on the debris concentration in basal ice. *Annals of Glaciology* **2**, 23–28.

Hambrey, M.J., Barrett, P.J., Ehrmann, W.U. and Larsen, B. (1992) Cenozoic sedimentary processes on the Antarctic continental margin and the record from deep drilling. *Zeitschrift für Geomorphologie* **Suppl. 86**, 77–103.

Hambrey, M.J., Larsen, B., Ehrmann, W.U. and Shipboard Scientific Party (1989) Forty million years of Antarctic glacial history revealed by Leg 119 of the Ocean Drilling Program. *Polar Record* **25**, 99–106.

Hambrey, M.J. and Müller, F. (1978) Structures and ice deformation in White Glacier, Axel Heiberg Island, NWT, Canada. *Journal of Glaciology* **20**, 41–67.

Hammer, C.U., Clausen, H.B. and Dansgaard, W. (1980) Greenland ice sheet evidence of post-glacial volcanism and its climate impact. *Nature* **288**, 230–235.

Handler, P. and Andsager, K. (1994) El Niño, volcanism and global climate. *Human Ecology* **22**, 37–57.

Hansen, J., Lacis, A., Rind, D., Russel, G., Stone, P., Fung, I., Ruedy, R. and Lerner, J. (1984) Climate sensitivity: analysis of feedback mechanisms. In Hansen, J.E. and Takahashi, T. (eds) *Climate Processes and Climatic Sensitivity*. Geophysical Monogram Series **29**, 130–163.

Harbor, J.M. (1992) Application of a general sliding law to simulating flow in a glacier cross-section. *Journal of Glaciology* **38**, 182–190.

Harmon, R.S., Mitterer, R.M., Kriausakal, N., Land, L.S., Schwarcz, H.P., Garrett, P., Larson, G.J., Vacher, H.L. and Rowe, M. (1983) U-series and amino-acid racemisation geochronology of Bermuda: implications for eustatic sea-level fluctuations over the past 250,000 years. *Palaeogeography, Palaeoclimatology, Palaeoecology* **44**, 41–70.

Harrison, S. and Winchester, V. (1998) Historical fluctuations of the Gualas and Reicher Glaciers, North Patagonian Icefield, Chile. *Holocene* **8**, 481–485.

Hays, J.D., Imbrie, J. and Shackleton, N.J. (1976) Variations in the Earth's orbit: pacemaker of the Ice Ages. *Science* **194**, 1121–1132.

Heinrich, H. (1988) Origin and consequences of cyclic ice-rafting in the Northeast Atlantic Ocean during the past 130,000 years. *Quaternary Research* **29**, 142–152.

Helmens, K.F., Rutter, N.W. and Kuhry, P. (1997) Glacier fluctuations in the eastern Andes of Columbia (South America) during the last 45,000 radiocarbon years. *Quaternary International* **38/39**, 39–48.

Herman, Y., Osmond, J.K. and Somayajulu, B.L.K. (1989) Late Neogene Arctic paleoceanography, micropaleontology, stable isotopes and chronology. In Herman, Y. (ed.) *The Arctic Seas: Climatology, Oceanography, Geology, and Biology*. Van Nostrand Reinhold, New York, 581–655.

Hessell, J.W.D. (1983) Climatic effects on the recession of the Franz Josef Glacier. *New Zealand Journal of Science* **26**, 315–320.

Heusser, C.J. (1989) Climate and chronology of Antarctica and adjacent South America over the past 30,000 yr. *Palaeogeography, Palaeoclimatology, Palaeoecology* **76**, 31–37.

Hillaire-Marcel, C. (1980) Multiple component postglacial emergence, eastern Hudson Bay, Canada. In Mörner, N.-A. (ed.) *Earth Rheology, Isostasy and Eustasy*. John Wiley, Chisester, 215–230.

Hodgson, D.A. (1985) The last glaciation of west-central Ellesmere Island, Arctic Archipelago, Canada. *Canadian Journal of Earth Sciences* **22**, 347–368.

Hodson, A.J., Tranter, M., Dowdeswell, J.A., Gurnell, A.M. and Hagen, J.O. (1997) Glacier thermal regime and suspended-sediment yield: a comparison of two high-Arctic glaciers. *Annals of Glaciology* **24**, 32–37.

Hoinkes, H.C. (1968) Glacier variation and weather. *Journal of Glaciology* **7**, 3–19.

Hollin, J.T. (1962) On the glacial history of Antarctica. *Journal of Glaciology* **4**, 173–195.

Holmgren, B. (1971) *Climate and Energy Exchange on a Sub-Polar Ice Cap in Summer, Arctic Institute of North America Devon Island Expedition 1961–1963*. Uppsala Universitet, Meteorologiska Institutionen Meddelande, Uppsala, 107–112.

Holmlund, P. (1997) Climatic influence on the size of glaciers in Northern Scandinavia during the last two centuries. In Frenzel, B., Boulton, G.S., Gläser, B. and Huckriede, U. (eds) *Glacier Fluctuations During the Holocene. Palaeoclimate Research* **24**, 115–124.

Holmlund, P., Karlén, W. and Grudd, H. (1996) Fifty years of mass balance and glacier front observations at the Tarfala Research Station. *Geografiska Annaler* **78A**, 105–114.

Holmlund, P. and Schneider, T. (1997) The effect of continentality on glacier response and mass balance. *Annals of Glaciology* **24**, 272–276.

Holmquist, B. and Wohlfarth, B. (1998) An evaluation of the Late Weichselian Swedish varve chronolgy based on cross-correlation analysis. *Geologiska Föreningen i Stockholms Förhandlingar* **120**, 35–46.

Holzhauser, H. (1984) Zur Geschichte der Aletschgletscher und des Fieschergletschers. *Physische Geographie* **13**, 1–448.

Holzhauser, H. (1997) Fluctuations of Grosser Aletch Glacier and the Gorner Glacier during the last 3200 years: new results. In Frenzel, B., Boulton, G.S., Gläser, B. and Huckriede, U. (eds) *Glacier Fluctuations During the Holocene. Palaeoclimate Research* **24**, 35–58.

Hooke, R. Le B., Calla, P., Holmlund, P., Nilsson, M. and Stroeven, A. (1989) A three-year record of seasonal variations in surface velocity, Storglaciären, Sweden. *Journal of Glaciology* **35**, 235–247.

Hormes, A., Stocker, T. and Schlüchter, C. (1998) 2000 yr cycles of Holocene climate variability stored underneath Unteraarglacier (Switzerland). *PAGES Open Science Meeting, London. Poster Abstracts*, 77.

Hughes, T. (1981) The weak underbelly of the West Antarctic ice sheet. *Journal of Glaciology* **27**, 518–525.

Hughes, T.J., Denton, G.H. and Fastook, J.L. (1985) The Antarctic Ice Sheet: an analog for northern hemisphere paleo-ice sheets? In Woldenberg, M.J. (ed.) *Models in Geomorphology*. Allen & Unwin, Boston, 25–72.

Hurrell, J.W. (1995) Decadal trends in the North Atlantic Oscillation: regional temperatures and precipitation. *Science* **269**, 676–679.

Hurrell, J.W. and van Loon, H. (1997) Decadal variations in climate associated with the North Atlantic Oscillation. *Climate Change* **36**, 301–326.

Hövermann, J. and Lehmkuhl, F. (1993) Bemerkungen zur eiszeitlichen Vergletscherungen Tibets. *Mitteilungen das Geographisches Geschellschaft. zu Lübeck* **58**, 137–158.

Imbrie, J., Berger, A., Boyle, E.A., Clemens, S.C., Duffy, A., Howard, W.R., Kukla, G., Kutzbach, J., Martinson, D.G., McIntyre, A., Mix, A.C., Molfino, B., Morley, J.J., Peterson, L.C., Pisias, N.G., Prell, W.L., Raymo, M.E., Shackleton, N.J. and Toggweiler, J.R. (1993a) On the structure and origin of major glaciation cycles. 2: The 100,000 year cycle. *Palaeoceanography* **8**, 699–735.

Imbrie, J., Berger, A. and Shackleton, N.J. (1993b) Role of orbital forcing: a two-million-year prespective. In Eddy, J.A. and Oeschger, H. (eds) *Global Changes in the Perspective of the Past*. John Wiley, Chichester, 263–277.

Imbrie, J., Boyle, E.A., Clemens, S.C., Duffy, A., Howard, W.R., Kukla, G., Kutzbach, J., Martinson, D.G., McIntyre, A., Mix, A.C., Molfino, B., Morley, J.J., Peterson, L.C., Pisias, N.G., Prell, W.L., Raymo, M.E., Shackleton, N.J. and Toggweiler, J.R. (1992) On the structure and origin of major glaciation cycles. 1: Linear responses to Milankovitch forcing. *Palaeoceanography* **7**, 701–738.

Imbrie, J.D., Hays, J.D., Martinson, D.G., McIntyre, A., Mix, A.C., Morley, J.J., Pisias, N.G., Prell, W.L. and Shackleton, N.J. (1984) The orbital theory of Pleistocene climate: support from revised chronology of the marine $\delta^{18}O$ record. In Berger, A., Imbrie, J., Hays, J., Kukla, G. and Saltzman (eds) *Milankovitch and Climate* **Part 1**, 269–305.

Imbrie, J. and Imbrie, J.Z. (1980) Modeling the climatic response to orbital variations. *Science* **207**, 943–953.

Imbrie, J. and Imbrie, K.P. (1979) *Ice Ages: Solving the Mystery*. Macmillan, London.

Indermühle, A., Stocker, T.F., Joos, F., Fischer, H., Smith, H.J., Wahlen, M., Deck, B., Mastroianni, D., Tschumi, J., Blunier, T., Meyer, R. and Stauffer, B. (1999) Holocene carbon-cycle dynamics based on CO_2 trapped in ice at Taylor Dome, Antarctica. *Nature* **398**, 121–126.

Ingolfsson, O. (1991) A review of the Late Weichselian and early Holocene glacial and environmental history of Iceland. In Maizels, J.K. and Caseldine, C. (eds) *Environmental Change in Iceland: Past and Present.* Kluwer Academic Publishers, Dordrecht, 13–29.

Ingolfsson, O., Björck, S., Haflidason, H. and Rundgren, M. (1997) Glacial and climatic events in Iceland reflecting regional North Atlantic climatic shifts during the Pleistocene–Holocene transition. *Quaternary Science Reviews* **16**, 1135–1144.

Innes, J.L. (1984) Relative dating of Neoglacial moraine ridges in North Norway. *Zeitschrift für Gletscherkunde und Glazialgeologie* **20**, 53–63.

Innes, J.L. (1985a) Lichenometry. *Progress in Physical Geography* **9**, 187–254.

Innes, J.L. (1985b) A standard *Rhizocarpon* nomenclature. *Boreas* **14**, 83–85.

Innes, J.L. (1986a) The use of percentage cover measurements in lichenometric dating. *Arctic and Alpine Research* **18**, 209–216.

Innes, J.L. (1986b) Influence of sampling design on lichen size–frequency distribution and its effect on derived lichenometric indices. *Arctic and Alpine Research* **18**, 201–208.

IPCC (1990) *Climate Change. The IPCC Scientific Assessment.* World Meteorological Organization/United Nations Environment Programme, Intergovernmental Panel on Climate Change. Cambridge University Press, Cambridge.

IPCC (1992) Climate Change: The Supplementary Report to the IPCC Scientific Assessment. Cambridge University Press, Cambridge.

IPCC (1995) *Climate Change 1995. The Science of Climate Change.* Contribution of Working Group I to the Second Assessment Report of the Intergovernmental Panel on Climate Change. Cambridge University Press, Cambridge.

Ives, J.D., Andrews, J.T. and Barry, R.G. (1975) Growth and decay of the Laurentide ice sheet and comparisons with Fenno-Scandia. *Naturwissenschaften* **62**, 118–125.

Jania, J. (1997) The problem of Holocene glacier and snow patches fluctuations in the Tatra Mountains: a short report. In Frenzel, B., Boulton, G.S., Gläser, B. and Huckriede, U. (eds) *Glacier Fluctuations During the Holocene. Palaeoclimate Research* **24**, 85–93.

Jania, J. and Hagen, J.O. (eds) (1996) Mass balance of Arctic glaciers. *IASC Report No. 5.*

Jansen, E. and Sjøholm, J. (1991) Reconstruction of glaciation over the past 6 million years from ice-borne deposits in the Norwegian Sea. *Nature* **349**, 600–604.

Johannesson, T., Raymond, C. and Waddington, E. (1989) Timescale for adjustment of glaciers to changes in mass balance. *Journal of Glaciology* **35**, 355–369.

Johannesson, T., Sigurdsson, O., Laumann, T. and Kennett, M. (1995) Degree–day glacier mass-balance modeling with applications to glaciers in Iceland, Norway and Greenland. *Journal of Glaciology* **41**, 345–358.

Johnsen, S.J., Clausen, H.B. Dansgaard, W., Fuhrer, K., Gundestrup, N., Hammer, C.U., Iversen, P., Jouzel, J., Stauffer, B. and Steffensen, J.P. (1992) Irregular glacial interstadials recorded in a new Greenland ice core. *Nature* **359**, 311–313.

Johnsen, S.J., Clausen, H.B., Dansgaard, W., Gundestrup, N.S., Hammer, C.U., Andersen, U., Andersen, K.K., Hvidberg, C.S., Dahl-Jensen, D., Steffensen, J.P., Shoji, H., Sveinbjörnsdóttir, A.E., White, J., Jouzel, J.

and Fisher, D. (1997) The δ^{18} O record along the Greenland Ice Core project deep ice core and problem of possible Eemian climatic instability. *Journal of Geophysical Research* **102** (C12), 26397–26410.

Johnsen, S.J., Dahl-Jensen, D., Dansgaard, W. and Gundestrup, N.S. (1995) Greenland temperatures derived from GRIP bore hole temperature and ice core isotope profiles. *Tellus* **B47**, 624–629.

Joussaume, S. (1989) Desert dust and climate: an investigation using a general circulation model. In Leinen, M. and Sarnthein, M. (eds) *Palaeoclimatology and Palaeometeorology: Modern and Past Patterns of Global Atmospheric Transport.* Kluwer, Dordrecht, 253–263.

Jouzel, J., Alley, R.B., Cuffey, K.M., Dansgaard, W., Grootes, P., Hoffmann, G., Johnsen, S.J., Koster, R.D., Peel, D., Shumann, C.A., Stievenard, M., Stuiver, M. and White, J. (1997) Validity of the temperature reconstruction from water isotopes in ice cores. *Journal of Geophysical Research* **102** (C12), 26471–26487.

Jouzel, J., Petit, J.R. and Raynaud, D. (1990) Palaeoclimatic information from ice cores: the Vostok record. *Transactions of the Royal Society of Edinburgh: Earth Sciences* **81**, 349–355.

Kaldal, I. and Vikingsson, S. (1991) Early Holocene deglaciation in central Iceland. *Jökull* **40**, 51–66.

Kamb, B. (1970) Sliding motion of glaciers: theory and observation. *Reviews of Geophysics and Space Physics* **8**, 673–728.

Karabanov, E., Prokopenko, A. and Williams, D. (1998) Intense glaciation of Siberia during Substage 5d: new evidence from Lake Baikal sediments. *PAGES Open Science Meeting, London. Poster Abstracts,* 81.

Karlén, W. (1976) Lacustrine sediments and tree-limit variations as indicators of Holocene climatic fluctuations in Lappland: Northern Sweden. *Geografiska Annaler* **58A**, 1–34.

Karlén, W. (1979) Glacier variations in the Svartisen area, northern Norway. *Geografiska Annaler* **61A**, 11–28.

Karlén, W. (1981) Lacustrine sediment studies. *Geografiska Annaler* **63A**, 273–281.

Karlén, W. (1988) Scandinavian glacial and climatic fluctuations during the Holocene. *Quaternary Science Reviews* 7, 199–209.

Karlén, W. (1998) Mount Kenyan glacier advances. *PAGES Open Science Meeting, London. Poster Abstracts,* p. 81.

Karlén, W. and Kuylenstierna, J. (1996) On solar forcing of Holocene climate: evidence from Scandinavia. *Holocene* **6**, 359–365.

Karlén, W. and Matthews, J.A. (1992) Reconstructing Holocene glacier variations from glacial lake sediments: studies from Nordvestlandet and Jostedalsbreen-Jotunheimen, southern Norway. *Geografiska Annaler* **74A**, 327–348.

Karlén, W., Fastook, J.L., Holmgren, K., Malmström, M., Matthews, J.A., Odada, E., Risberg, J., Rosquist, G., Sandgren, P., Shemesh, A. and Westerberg, L.-O. (1999) Glacier fluctuations on Mount Kenya since ~6000 cal. years BP: Implications for Holocene climatic change in Africa. *Ambio* **28**, 409–418.

Karpuz, N. and Jansen, E. (1992) A high-resolution diatom record of the last deglaciation from the SE Norwegian Sea: documentation of rapid climatic change. *Palaeoceanography* **7**, 499–520.

Karpuz, N., Jansen, E. and Haflidason, H. (1993) Palaeoceanographic reconstruction of surface ocean conditions in the Greenland, Iceland and Norwegian Seas through the last 14 ka based on diatoms. *Quaternary Science Reviews* **12**, 115–140.

Kaser, G. and Georges, C. (1997) Changes of the equilibrium-line altitude in the tropical Cordillera Blanca, Peru, 1930–50, and their spatial variations. *Annals of Glaciology* **24**, 344–349.

Kaufmann, G. and Lambeck, K. (1997) Implications of Late Pleistocene glaciation of the Tibetan Plateau for present-day uplift rates and gravity anomalies. *Quaternary Research* **48**, 267–279.

Keeling, C.D. and Whorf, T.P. (1998) *Atmospheric CO_2 Concentrations (ppmv) Derived from in situ Air Samples Collected at Mauna Loa Observatory, Hawaii.* Scripps Institute of Oceanography, La Jolla, California.

Keeling, C.D., Whorf, T.P., Wahlen, M. and Van der Plicht, J. (1995) Interannual extremes in the rate of rise of atmospheric carbon dioxide since 1980. *Nature* **375**, 666–670.

Kelly, P.M. and Wigley, T.M.L. (1990) The influence of solar forcing trends on global mean temperature since 1861. *Nature* **347**, 460–462.

Kelly, P.M. and Sear, C.B. (1984) Climate impact of explosive volcanic eruptions. *Nature* **311**, 740–743.

Kemp, D.D. (1994) *Global Environmental Issues – A Climatological Approach.* Routledge, London.

Kennett, M. and Eiken, T. (1997) Airborne measurement of glacier surface elevation by scanning laser altimeter. *Annals of Glaciology* **24**, 293–296.

Kerschner, H. (1997) Statistical modelling of equilibrium-line altitudes of Hintereisferner, central Alps, Austria, 1859–present. *Annals of Glaciology* **24**, 111–115.

Kerschner, H., Ivy-Ochs, S., Sailer, R. and Schlüchter, C. (1998) Some paleoclimatic implications from Younger Dryas glaciers in the Alps. *PAGES Open Science Meeting, London. Poster Abstracts*, 82.

Kieffer, H.H., Kargel, J.S., Lucchitta, B.K., Williams, R.C. and Hall, D. (1994) Opportunities for glacier monitoring using the AM-1 satellite of the Earth Observing System (1998 launch). *NSIDC Notes* **10**, 5–6.

Kirkbride, M.P. and Brazier, V. (1998) A critical evaluation of the use of glacier chronologies in climate reconstruction, with reference to New Zealand. *Quaternary Proceedings* **6**, 55–64.

Kirkbride, M.P. and Warren, C.R. (1997) Calving processes at a grounded ice cliff. *Annals of Glaciology* **24**, 116–121.

Kjøllmoen, B. (1998) *Glasiologiske Undersøkelser i Norge 1996 og 1997. Rapport 20*, Norges vassdrags-og energiverk, Oslo.

Klassen, R.A. and Fisher, D.A. (1988) Basal flow conditions at the northeastern margin of the Laurentide ice sheet, Lanchaster Sound. *Canadian Journal of Earth Sciences* **25**, 1740–1750.

Koch, L. (1945) The East Greenland ice. *Meddelelser om Grønland* **130**, no. 3.

Konzelmann, T. and Braithwaite, R.J. (1995) Variations of ablation, albedo and energy-balance at the margin of the Greenland ice sheet, Kronsprins-Chriastian Land, eastern north Greenland. *Journal of Glaciology* **41**, 174–182.

Kromer, B. and Becker, B. (1993) German oak and pine ^{14}C calibration 7200–9400 B.C. *Radiocarbon* **35**, 125–137.

Kruss, P. (1983) Climate change in East Africa: numerical simulation from the 100 years of terminus record at Lewis Glacier, Mount Kenya. *Zeitschrift für Gletscherkunde und Glazialgeologie* **19**, 43–60.

Kugelmann, O. (1991) Dating recent glacier advances in the Svarfadardalur-Skídadalur area of northern Iceland by means of a new lichen curve. In Maizels, J.K. and Caseldine, C. (eds) *Environmental Change in Iceland: Past and Present.* Kluwer Academic Publishers, Dordrecht, 203–217.

Kuhle, M. (1987) The problem of a Pleistocene Inland Glaciation of the Northeastern Qinghai-Xizang (Tibet) Plateau. In Hövermann, J. and Wang, W. (eds) *Reports of the Qinghai-Xizang (Tibet) Plateau.* Beijing, 250–315.

Kuhle, M. (1988) Geomorphological findings on the build-up of Pleistocene glaciation in Southern Tibet and on the problem of inland ice. *GeoJournal* **17**, 457–512.

Kuhle, M. (1991) Observations supporting the Pleistocene inland glaciation of High Asia. *GeoJournal* **25**, 131–231.

Kuhle, M. (1998) Reconstruction of the 2.4 million km^2 late Pleistocene ice sheet on the Tibetan Plateau and its impact on the global climate. *Quaternary International* **45**, 71–108.

Kuhle, M., Herterich, K. and Calov, R. (1989) On the ice age glaciation of the Tibetan Highlands and its transformation into a 3-D model. *GeoJournal* **19**, 201–206.

Kuhn, M. (1979) On the computation of heat transfer coefficients from energy-balance gradients on a glacier. *Journal of Glaciology* **22**, 263–272.

Kuhn, M., Schlosser, E. and Span, N. (1997) Eastern Alpine glacier activity and climatic records since 1860. *Annals of Glaciology* **24**, 164–168.

Kukla, G.J. (1977) Pleistocene land-sea correlations. *Earth Science Reviews* **13**, 307–374.

Kukla, G.J. (1987) Pleistocene climates in China and Europe compared to oxygen isotope record. *Palaeoecology of Africa* **18**, 37–45.

Kukla, G.J. (1989) Long continental records of climate – an introduction. *Palaeogeography, Palaeoclimatology, Palaeoecology* **72**, 1–9.

Kukla, G.J. and Gavin, J. (1992) Insolation regime of the warm to cold transitions. In Kukla, G.J. and Went, E. (eds) *Start of a Glacial.* Springer-Verlag, Berlin, 307–339.

Kutzbach, J.E. and Ruddiman, W.F. (1993) Model description, external forcing and surface boundary conditions. In Wright, H.E. Jr, Kutzbach, J.E., Webb III, T., Ruddiman, W.F., Street-Perrott, F.A. and Bartlein, P.J. (eds) *Global Climates Since the Last Glacial Maximum.* University of Minnesota Press, Minneapolis, 12–23.

Kutzbach, J.E. and Webb III, T. (1993) Conceptual basis for understanding Late Quaternary climates. In Wright, H.E. Jr, Kutzbach, J.E., Webb III, T., Ruddiman, W.F., Street-Perrott, F.A. and Bartlein, P.J. (eds) *Global Climates Since the Last Glacial Maximum.* University of Minnesota Press, Minneapolis, 5–11.

Kutzbach, J.E. and Wright, H.E. (1985) Simulation of the climate of 18,000 yr BP: results for the North American/European sector and comparison with the geologic record. *Quaternary Science Reviews* **4**, 147–187.

Labeyrie, L.D., Pichon, J.J., Labracherie, M., Ippolito, P., Duprat, J. and Duplessy, J.C. (1986) Melting history of Antarctica during the past 60,000 years. *Nature* **322**, 701–706.

LaMarche, V.C. Jr and Hirschboeck, K.K. (1984) Frost rings in trees as records of major volcanic eruptions. *Nature* **307**, 121–126.

Lamb, H.H. (1970) Volcanic dust in the atmosphere: with a chronology and assessment of its meteorological significance. *Philosophical Transactions of the Royal Society* **A226**, 425–533.

Lambeck, K. (1991a) A model for Devensian and Flandrian glacial rebound and relative sea-level change in Scotland. In Sabadini, R., Lambeck, K. and Boschi, E. (eds) *Glacial Isostasy, Sea Level and Mantle Rheology.* Kluwer, Dordrecht, 33–62.

Lambeck, K. (1991b) Glacial rebound and sea-level change in the British Isles. *Terra Nova* **3**, 379–389.

Lambeck, K. (1993a) Glacial rebound of the British Isles. I: Preliminary model results. *Geophysical Journal International* 115, 941–959.

Lambeck, K. (1993b) Glacial rebound of the British Isles. II. A high-resolution, high-precision model. *Geophysical Journal International* **115**, 960–990.

Landvik, J.Y., Bondevik, S., Elverhøi, A., Fjeldskaar, W., Mangerud, J., Siegert, M.J., Salvigsen, O., Svendsen, J.I. and Vorren, T.O. (1998) The last glacial maximum of Svalbard and the Barents Sea area: ice sheet extent and configuration. *Quaternary Science Reviews* **17**, 43–75.

Larsen, E., Funder, S. and Thiede, J. (1999a) Late Quaternary history of northern Russia and adjacent shelves – a synopsis. *Boreas* **28**, 6–11.

Larsen, E., Lyså, A., Demidov, I., Funder, S., Houmark-Nielsen, M., Kjær, K.H. and Murray, A.S. (1999b) Age and extent of the Scandinavian ice sheet in northwest Russia. *Boreas* **28**, 115–132.

Laumann, T. and Reeh, N. (1993) Sensitivity to climate change of the mass balance of glaciers in southern Norway. *Journal of Glaciology* **39**, 656–665.

Lautenschlager, M. and Herterich, H. (1990) Atmospheric response to ice age conditions: climatology near the Earth's surface. *Journal of Geophysical Research* **95**, 2547–2557.

Lautenschlager, M., Mikolajewicz, U., Meier-Reimer, E. and Heinze, C. (1992) Application of ocean models for the interpretation of atmospheric global circulation model experiments on the climate of the last glacial maximum. *Paleoceanography* **7**, 769–782.

Lawson, W. (1997) Spatial, temporal and kinematic characteristics of surges of Variegated Glacier, Alaska. *Annals of Glaciology* **24**, 95–101.

Leemann, A. and Niessen, F. (1994) Holocene glacial activity and climatic variations in the Swiss Alps: reconstructing a continuous record from proglacial lake sediments. *Holocene* **4**, 259–268.

Lehman, S.J. and Keigwin, L.D. (1992) Sudden changes in North Atlantic circulation during the last deglaciation. *Nature* **356**, 757–762.

Lehmkuhl, F. (1994) Geomorphologische Untersuchungen zum Klima des Holozäns und Jungpleistozäns Osttibets. *Göttinger Geographische Abhandlungen* **102**, 1–184.

Lehmkuhl, F. (1997) Late Pleistocene, Late-Glacial and Holocene glacier advances on the Tibetan Plateau. *Quaternary International* **38/39**, 77–83.

Lehmkuhl, F. (1998) Quaternary glaciations in central and western Mongolia. *Quaternary Proceedings* **6**, 153–167.

Lehmkuhl, F., Owen, L.A. and Derbyshire, E. (1998) Late Quaternary glacial history of northeast Tibet. *Quaternary Proceedings* **6**, 121–142.

Leonard, E.M. (1986) Use of lacustrine sedimentary sequences as indicators of Holocene glacial history, Banff National Park, Alberta, Canada. *Quaternary Research* **26**, 218–231.

Letreguilly, A. (1988) Relation between the mass balance of western Canadian mountain glaciers and meteorological data. *Journal of Glaciology* **34**, 11–18.

Levesque, A.J., Mayle, F.E. and Walker, I.R. (1993) The Amphi-Atlantic Oscillation: a proposed Late-glacial climatic event. *Quaternary Science Reviews* **12**, 629–643.

Licciardi, J.M., Clark, P.U., Jensons, J.W. and Macayeal, D.R. (1998) Deglaciation of a soft-bedded Laurentide ice sheet. *Quaternary Science Reviews* **17**, 427–448.

Liestøl, O. (1967) Storbeen glacier in Jotunheimen, Norway. *Norsk Polarinstitutt Skrifter* **14**, 1–63.

Loewe, F. (1971) Considerations of the origin of the Quaternary ice sheet in North America. *Arctic and Alpine Research* **3**, 331–344.

Lorius, C., Barkov, N.I., Jouzel, J., Korotkevitch, Y.S., Kotlyakov, V.M. and Raynaud, D. (1988) Antarctic ice core: CO_2 and climate change over the last climate cycle. *EOS* **June 28**, 681–684.

Lorius, C., Jouzel, J., Raynaud, D., Hansen, J. and le Treut, H. (1990) The ice-core rcord: climatic sensitivity and future greenhouse warming. *Nature* **347**, 139–145.

Lorius, C., Raisbeck, G., Jouzel, J. and Raynaud, D. (1989) Long-term environmental records from Antarctic ice cores. In Oeschger, F. and Langway, C.C. Jr (eds) *The Environmental Record in Glaciers and Ice Sheets.* John Wiley, Chichester, 343–362.

Lotter, A.F. (1991) Absolute dating of the Late-Glacial period in Switzerland using annually laminated sediments. *Quaternary Research* **35**, 321–330.

Loutre, M.F. (1993) *Evolution de Línlandsis Groenlandais au Cours des 5000 Prochaines Années.* Scientific Report 1993/9. Institut d'Astronomie et de Géopysique G. Lemaitre, Université Catholique de Lovain, Lovain-la-Neuve.

Lowe, J.J. and Gray, J.M. (1980) The stratigraphic subdivision of the Lateglacial of northwest Europe. In Lowe, J.J., Gray, J.M. and Robinson, J.E. (eds) *Studies in the Lateglacial of North-West Europe.* Pergamon, Oxford, 157–175.

Lowe, J.J. and Members of NASP (1995) Paleoclimate of the North Atlantic seaboards during the last glacial/interglacial transition. *Quaternary International* **28**, 51–61.

Lowe, J.J. and Walker, M.J.C. (1997) *Reconstructing Quaternary Environments.* Longman, New York.

Lowell, T.V., Heusser, C.J. and Andersen, B.G. (1995) Interhemispheric correlation of Late Pleistocene glacial events. *Science* **269**, 1541–1549.

Luckman, B. (1986) Reconstruction of Little Ice Age events in the Canadian Rocky Mountains. *Géographie Physique et Quaternaire* **40**, 17–28.

Luckman, B.H., Holdsworth, G and Osborn, G.D. (1993) Neoglacial glacier fluctuations in the Canadian Rockies.*Quaternary Research* **39**, 144–153.

Lundquist, J. (1986) Late Weichselian glaciation and deglaciation in Scandinavia. *Quaternary Science Reviews* **5**, 269–292.

Maggi, V. (1997) Mineralogy of atmospheric microparticles deposited along the greenland Ice Core Project ice core. *Journal of Geophysical Research* **102** (C12), 26725–26734.

Magny, M. (1993) Solar influences on Holocene climate changes, illustrated by correlation between past lake level fluctuations and the atmospheric ^{14}C record. *Quaternary Research* **40**, 1–9.

Mahaney, W.C. (1987) Lichen trimlines and weathering features as indicators of mass balance changes and successive retreat stages of the Mer de Glace in the western Alps. *Zeitschrift für Geomorphologie* **31**, 411–418.

Mahaney, W.C. (1988) Holocene glaciations and paleoclimate of Mount Kenya and other East African mountains. *Quaternary Science Reviews* **7**, 211–225.

Mangerud, J. (1989) Correlation of the Eemian and the Weichselian with deep sea oxygen isotope stratigraphy. *Quaternary International* **3/4**, 1–4.

Mangerud, J. (1991) The last interglacial/glacial cycle in northern Europe. In Shane, L.C.K. and Cushing, E.J. (eds) *Quaternary Landscapes*. University of Minnesota Press, Minneapolis, 38–75.

Mangerud, J., Andersen, S.T., Berglund, B. and Donner, J.J. (1974) Quaternary stratigraphy of Norden, a proposal for teminology and classification. *Boreas* **3**, 109–128.

Mangerud, J., Bolstad, M., Elgersma, A., Helliksen, D., Landvik, J.Y., Lønne, I., Lycke, A.K., Salvigsen, O., Sandahl, T. and Svendsen, J.I. (1992) The last glacial maximum in Spitsbergen, Svalbard. *Quaternary Research* **38**, 633–664.

Mangerud, J., Dokken, T.M., Hebbeln, D., Heggen, B., Ingolfsson, O., Landvik, J.Y., Mejdahl, V., Svendsen, J.I. and Vorren, T.O. (1998) Fluctuations of the Svalbard-Barents Sea ice sheet during the last 150 000 years. *Quaternary Science Reviews* **17**, 11–42.

Mangerud, J., Jansen, E. and Landvik, J. (1996) Late Cenozoic history of the Scandinavian and Barents Sea ice sheets. *Global and Planetary Change* **12**, 11–26.

Mangerud, J. and Svendsen, J.I. (1990) Deglaciation chronology inferred from marine sediments in a proglacial lake basin, western Spitsbergen, Svalbard. *Boreas* **19**, 249–272.

Mangerud, J. and Svendsen, J.I. (1992) The last interglacial–glacial period on Spitsbergen, Svalbard. *Quaternary Science Reviews* **11**, 633–664.

Mangerud, J., Svendsen, J.I. and Astakhov, V. (1999) Age and extent of the Barents and Kara ice sheets in Northern Russia. *Boreas* **28**, 46–80.

Marden, C.J. (1997) Late-Glacial fluctuations of south Patagonian icefield, Torres del Paine National Park, southern Chile. *Quaternary International* **38/39**, 61–68.

Martinson, D.G., Pisias, N.G., Hays, J.D., Imbrie, J., Moore, T.C. and Shackleton, N.J. (1987) Age dating and the orbital theory of ice ages: development of a high resolution 0–300,000 year chronostratigraphy. *Quaternary Research* **27**, 1–29.

Maslanik, J.A., Serreze, M.C. and Barry, R.G. (1996) Recent decreases in Arctic summer ice cover and linkages to atmospheric circulation anomalies. *Geophysical Research Letters* **23**, 1677–1680.

Maslin, M.A., Li, X.S., Loutre, M.-F. and Berger, A. (1998) The contribution of orbital forcing to the progressive intensification of northern hemisphere glaciation. *Quaternary Science Reviews* **17**, 411–426.

Mass, C. and Schneider, S.H. (1977) Influence of sunspots and volcanic dust on long-term temperature records inferred by statistical investigations. *Journal of Atmospheric Sciences* **34**, 1195–2204.

Matthews, J.A. (1977) A lichenometric test of the 1750 end-moraine hypothesis: Storbreen gletschervorfeld, southern Norway. *Norsk Geografisk Tidsskrift* **31**, 129–136.

Matthews, J.A. (1985) Radiocarbon dating of surface and buried soils: principles, problems and prospects. In Richards, K.S., Arnett, R.R. and Ellis, S. (eds) *Geomorphology and Soils*. George Allen & Unwin, London, 269–288.

Matthews, J.A. (1991) The late Neoglacial ('Little Ice Age') glacier maximum in southern Norway: new ^{14}C-dating evidence and climatic implications. *Holocene* **1**, 219–233.

Matthews, J.A. (1992) *The Ecology of Recently Deglaciated Terrain*. Cambridge University Press, Cambridge.

Matthews, J.A. (1997) Dating problems in the investigation of Scandinavian Holocene glacier variations. In Frenzel, B., Boulton, G.S., Gläser, B. and Huckriede, U. (eds) *Glacier Fluctuations During the Holocene. Palaeoclimate Research* **24**, 141–157.

Matthews, J.A. and Caseldine, C.J. (1987) Arctic-alpine brown soils as a source of palaeoenvironmental information: further ^{14}C dating and palynological evidence from Vestre Memurubreen, Jotunheimen, Norway. *Journal of Quaternary Science* **2**, 59–71.

Matthews, J.A. and Dresser, P.Q. (1983) Intensive ^{14}C dating of a buried palaeosol horizon. *Geologiska Föreningen i Stockholm, Förhandlingar* **105**, 59–63.

Matthews, J.A. and Karlén, W. (1992) Asynchronous neoglaciation and Holocene climatic change reconstructed from Norwegian glaciolacustrine sedimentary sequences. *Geology* **20**, 991–994.

Matthews, J.A., Nesje, A. and Dahl, S.O. (1996) Reassessment of supposed early 'Little Ice Age' and older Neoglacial moraines in the Sandane area of western Norway. *Holocene* **6**, 106–110.

Matthews, J.A. and Shakesby, R. A. (1984) The status of the 'Little Ice Age' in southern Norway: relative-age dating of Neoglacial moraines with Schmidt hammer and lichenometry. *Boreas* **13**, 333–346.

Mayewski, P.A., Lyons, W.B., Spencer, M.J., Twickler, M.S., Buck, C.F. and Whitlow, S. (1990) An ice-core record of atmospheric response to anthropogenic sulphate and nitrate. *Nature* **346**, 554–556.

Mayewski, P.A., Meeker, L.D., Twickler, M.S., Whitlow, S., Yang, Q., Lyons, W.B. and Prentice, M. (1997) Major features and forcing of high-latitude northern hemisphere atmospheric circulation using a 110,000-year-long glaciochemical series. *Journal of Geophysical Research* **102** (C12), 26345–26366.

McCabe, G.J. and Fountain, A.G. (1995) Relations between atmospheric circulation and mass balance of South Cascade Glacier, Washington, USA. *Arctic and Alpine Research* **27**, 226–233.

McCarroll, D. (1989) Potential and limitations of the Schmidt hammer for relative-age dating: field tests on Neoglacial moraines, Jotunheimen, southern Norway. *Arctic and Alpine Research* **21**, 268–275.

McCarroll, D. (1994) A new approach to lichenometry: dating single-age and diachronous surfaces. *Holocene* **4**, 383–396.

McCarroll, D. and Nesje, A. (1993) Vertical extent of ice sheets in Nordfjord, western Norway: measuring degree of rock surface weathering using Schmidt hammer and rock surface roughness. *Boreas* **22**, 255–265.

McClung, D.M. and Armstrong, R.L. (1993) Temperate glacier time response from field data. *Journal of Glaciology* **39**, 323–326.

McManus, J.F., Bond, G.C., Broecker, W.S., Johnsen, S., Labeyrie, L. and Higgins, S. (1994) High-resolution climatic records from the North Atlantic during the last interglacial. *Nature* **371**, 326–329.

Meeker, L.D., Mayewski, P.A., Twickler, M.S., Whit-low, S.I. and Meese, D. (1997) A 110,000-year history of change in continental biogenic emission and related atmospheric circulation inferred from the Greenland Ice Sheet Project Ice Core. *Journal of Geophysical Research* **102** (C12), 26489–26504.

Meese, D.A., Gow, A.J., Alley, R.B., Zielinski, G.A., Grootes, P.M., Ram, M., Taylor, K.C., Mayewski, P.A. and Bolzan, J.F. (1997) The Greenland Ice Sheet Project 2 depth–age scale: methods and results. *Journal of Geophysical Research* **102** (C12), 26411–26423.

Meier, M.F. (1984) Contribution of small glaciers to global sea level. *Science* **226**, 1418–1421.

Menounos, B. and Reasoner, M.A. (1997) Evidence for cirque glaciation in the Colorado Front Range during the Younger Dryas chronozone. *Quaternary Research* **48**, 38–47.

Menzies, J. (ed.) (1995) *Modern Glacial Environments*. Butterworth-Heinemann, Oxford.

Mercer, J.H. (1978) West Antarctic ice sheet and CO_2 greenhouse effect: a threat of disaster. *Nature* **271**, 321–325.

Mercer, J.H. (1983) Cenozoic glaciation in the Southern Hemisphere. *Annual Review of Earth and Planetary Science* **11**, 99–132.

Mercer, J.H. and Palacios, O. (1977) Radiocarbon dating of the last glaciation in Peru. *Geology* **5**, 600–604.

Mikhalenko, V.N. (1997) Changes in Eurasian glaciation during the past century: glacier mass balance and ice-core evidence. *Annals of Glaciology* **24**, 283–287.

Miles, M.K. and Gildersleeves, P.B. (1978) Volcanic dust and changes in Northern Hemisphere temperatures. *Nature* **271**, 735–736.

Miller, G.H., Sejrup, H.P., Lehman, S.J. and Forman, S.L. (1989) Glacial history and marine environmental change during the last interglacial–glacial cycle, western Spitsbergen, Svalbard. *Boreas* **18**, 273–296.

Miller, G.H. and Vernal, A. (1992) Will greenhouse warming lead to Northern Hemisphere ice-sheet growth? *Nature* **355**, 244–246.

Miller, S.L. (1969) Clathrate hydrates of air in Antarctic ice. *Science* **165**, 489–490.

Möller, P., Bolshiyanov, D.Y. and Bergsten, H. (1999) Weichselian geology and palaeoenvironmental history of the central Taymyr Peninsula, Siberia, indicating no glaciation during the last glacial maximum. *Boreas* **28**, 92–114.

Mörner, N.-A. (1993) Global changes: the last millennia. *Global and Planetary Change* **7**, 211–217.

Motyka, R.J. and Begét, J.E. (1996) Taku Glacier southeast Alaska, USA: Late Holocene history of a tidewater glacier. *Arctic and Alpine Research* **28**, 42–51.

Murray, T., Gooch, D.L. and Stuart, G.W. (1997) Structures within the surge front at Bakaninbreen, Svalbard, using ground penetrating radar. *Annals of Glaciology* **24**, 122–129.

Naruse, R., Skvarca, P. and Takeuchi, Y. (1997) Thinning and retreat of Glacier Uppsala, and an estimate of annual ablation changes in southern Patagonia. *Annals of Glaciology* **24**, 38–42.

Neftel, A., Oeschger, H., Schwander, J., Stauffer, B. and Zumbrunn, R. (1982) Ice core sample measurements give atmospheric CO_2 content during the past 40,000 years. *Nature* **295**, 220–223.

Neftel, A., Oeschger, H., Staffelbach, T. and Stauffer, B. (1988) CO_2 record in the Byrd ice core 50,000–5,000 years BP. *Nature* **331**, 609–611.

Nesje, A. (1992) Topographical effects on the equilibrium-line altitude on glaciers. *GeoJournal* **27.4**, 383–391.

Nesje, A. (1994) A gloomy 250-year memory; the glacier destruction of the Tungøyane farm in Oldedalen, western Norway, 12 December 1743. *Norsk Geografisk Tidsskrift* **48**, 133–135.

Nesje, A. and Dahl, S.O. (1990) Autochthonous blockfields in southern Norway: implications for the geometry, thickness and isostatic loading of the late Weichselian Scandinavian ice sheet. *Journal of Quaternary Science* **5**, 225–234.

Nesje, A. and Dahl, S.O. (1991a) Holocene glacier variations of Blåisen, Hardangerjøkulen, central southern Norway. *Quaternary Research* **35**, 25–40.

Nesje, A. and Dahl, S.O. (1991b) Late Holocene glacier fluctuations in Bevringsdalen, Jostedalsbreen region, western Norway (ca. 3200–1400 BP). *Holocene* **1**, 1–7.

Nesje, A. and Dahl, S.O. (1992) Geometry, thickness and isostatic loading of the Late Weichselian Scandinavian ice sheet. *Norsk Geologisk Tidsskrift* **72**, 271–273.

Nesje, A., Dahl, S.O., Anda, E. and Rye, N. (1988) Blockfields in southern Norway: significance for the late Weichselian ice sheet. *Norsk Geologisk Tidsskrift* **68**, 149–169.

Nesje, A., Johannessen, T. and Birks, H.J.B. (1995) Briksdalsbreen, western Norway: climatic effects on the terminal response of a temperate glacier between AD 1901 and 1994. *Holocene* **5**, 343–347.

Nesje, A. and Kvamme, M. (1991) Holocene glacier and climate variations in western Norway: evidence for early Holocene glacier demise and multiple Neoglacial events. *Geology* **19**, 610–612.

Nesje, A., Kvamme, M., Rye, N. and Løvlie, R. (1991) Holocene glacial and climate history of the Jostedalsbreen region, western Norway; evidence from lake sediments and terrestrial deposits. *Quaternary Science Reviews* **10**, 87–114.

Nesje, A. and Sejrup, H.-P. (1988) Late Weichselian/Devensian ice sheets in the North Sea and adjacent land areas. *Boreas* **17**, 371–384.

Norddahl, H. (1990) Late Weichselian and early Holocene deglaciation of Iceland. *Jökull* **40**, 27–50.

Nuttall, A.-M., Hagen, J.O. and Dowdeswell, J. (1997) Quiescent-phase changes in velocity and geometry of Finsterwalderbreen, a surge-type glacier in Svalbard. *Annals of Glaciology* **24**, 249–254.

Nye, J.F. (1957) The distribution of stress and velocity in glaciers and ice sheets. *Proceedings of the Royal Society of London Series A* **239**, 113–133.

Nye, J.F. (1958) Surges in glaciers. *Nature* **181**, 1450–1451.

Nye, J.F. (1960) The response of glaciers and ice sheets to seasonal and climatic changes. *Proceedings of the Royal Society of London Series A* **256**, 559–584.

Nye, J. (1976) Water flow in glaciers: jökulhlaups, tunnels and veins. *Journal of Glaciology* **17**, 181–207.

Oerlemans, J. (1988) Simulation of historic glacier variations with a simple climate–glacier model. *Journal of Glaciology* **34**, 333–341.

Oerlemans, J. (ed.) (1989) *Glacier Fluctuations and Climate Change*. Kluwer Academic Publishers, Dordrecht.

Oerlemans, J. (1991) The mass balance of the Greenland ice sheet: sensitivity to climate change as revealed by energy-balance modelling. *Holocene* **1**, 40–49.

Oerlemans, J. (1992) Climate sensitivity of glaciers in southern Norway: application of an energy-balance method to Nigardsbreen, Hellstugubreen and Ålfotbreen. *Journal of Glaciology* **38**, 223–232.

Oerlemans, J. (1994) Quantifying global warming from the retreat of glaciers. *Science* **264**, 243–245.

Oerlemans, J. (1997) A flowline model for Nigardsbreen, Norway: projection of future glacier length based on dynamic calibration with the historic record. *Annals of Glaciology* **24**, 382–389.

Oerlemans, J. and Fortuin, J.P.F. (1992) Sensitivity of glaciers and small ice caps to greenhouse warming. *Science* **258**, 115–118.

Oerlemans, J. and Hoogendoorn, N.C. (1989) Mass balance gradients and climatic change. *Journal of Glaciology* **35**, 399–405.

Oeschger, H. (1992) Working hypothesis for glaciation/ deglaciation mechanisms. In Bard, E. and Broecker, W.S. (eds) *The Last Deglaciation: Absolute and Radiocarbon Chronologies. NATO ASI Series, 1,2.* Springer-Verlag, Berlin, 273–289.

Ogilvie, A.E.J. (1984) The past climate and sea-ice record from Iceland. *Climate Change* **6**, 131–152.

Ogilvie, A.E.J. (1998) Historical accounts of weather event, sea ice and related matters in Iceland and Greenland, AD c. 1250–1430. In Wishman, E., Frenzel, B. and Weiss, M.M. (eds) Documentary climatic evidence for 1750–1850 and the fourteenth century. *Palaeoclimate Research* **23**, 25–43.

Ohmura, A., Kasser, P. and Funk, M. (1992) Climate at the equilibrium line of glaciers. *Journal of Glaciology* **38**, 397–411.

Ommanney, C.S.L. (1969) *A Study of Glacier Inventory: the Ice Masses of Axel Heiberg Island, Canadian Arctic Archipelago.* Axel Heiberg Research Report, Glaciology **3**, McGill University.

Orombelli, G. and Mason, P. (1997) Holocene glacier fluctuations in the Italian alpine region. In Frenzel, B., Boulton, G.S., Gläser, B. and Huckriede, U. (eds) *Glacier Fluctuations During the Holocene. Palaeoclimate Research* **24**, 57–65.

Osborn, G. (1985) Holocene tephrostratigraphy and glacier fluctuations in Waterton Lakes and Glacier National parks, Alberta and Montana. *Canadian Journal of Earth Sciences* **22**, 1093–1101.

Osborn, G. and Gerloff, L. (1997) Latest Pleistocene and early Holocene fluctuations of glaciers in the Canadian and northern American Rockies. *Quaternary International* **38/39**, 7–19.

Osborn, G. and Luckman, B. (1988) Holocene glacier fluctuations in the Canadian Cordillera (Alberta and British Colombia). *Quaternary Science Reviews* **7**, 115–128.

Overpeck, J., Hughen, K., Hardy, D., Bradley, R., Case, R., Douglas, M., Finney, B., Gajewski, K., Jacoby, G., Jennings, A., Lamoureux, S., Lasca, A., MacDonald, G., Moore, J., Retelle, M., Smith, S., Wolfe, A. and Zielinski, G. (1997) Arctic environmental change of the last four centuries. *Science* **278**, 1251–1256.

Owen, L.A., Derbyshire, E. and Fort, M. (1998) The Quaternary glacial history of the Himalaya. *Quaternary Proceedings* **6**, 91–120.

Paterson, W.S.B. (1994) *The Physics of Glaciers* (3rd edn). Elsevier Science, Amsterdam.

Pelfini, M. and Smiraglia, C. (1997) Signals of 20th-century warming from the glaciers in the Central Italian

Alps. *Annals of Glaciology* **24**, 350–354.

Peltier, W.R. (1987) Mechanisms of relative sea-level change and the geophysical responses to ice-water loading. In Devoy, R.J.N. (ed.) *Sea Surface Studies.* Croom Helm, London, 57–94.

Pelto, M.S. (1988) The annual balance of North Cascade glaciers, Washington, USA, measured and predicted using an activity-index method. *Journal of Glaciology* **34**, 194–199.

Pelto, M.S. (1989) Time-series analysis of mass balance and local climatic records from four northwestern North American glaciers. In Colbeck, S.C. (ed.) *Snow Cover and Glacier Variations.* IAHS Publication 183, Great Yarmouth, UK, 95–102.

Penck, A. and Brückner, E. (1909) *Die Alpen im Eiszeitalter.* Tauchnitz, Leipzig.

Perkins, J.A. and Sims, J.D. (1983) Correlation of Alaskan varve thickness with climatic parameters, and the use in palaeoclimatic reconstruction. *Quaternary Research* **20**, 308–321.

Pestieux, P., Duplessy, J.-C. and Berger, A. (1987) Palaeoclimatic variability at frequencies ranging from 10^{-4} cycle per year to 10^{-3} cycle per year – evidence for nonlinear behaviour of the climate system. In Rampino, M.R., Sanders, J.E., Newman, W.S. and Königsson, L.K. (eds) *Climate, History, Periodicity and Predictability.* Van Nostrand Reinhold, New York, 285–299.

Petit, J.R., Jouzel, J., Raynaud, D., Barkov, N.I., Barnola, J.-M., Basile, I., Bender, M., Chappellaz, J., Davis, M., Delaygue, G., Delmotte, M., Kotlyakov, V.M., Legrand, M., Lipenkov, V.Y., Lorius, C., Pépin, L., Ritz, C., Saltzman, E. and Stievenard, M. (1999) Climate and atmospheric history of the past 420,000 years from the Vostok ice core, Antarctica. *Nature* **399**, 429–436.

Pirazzoli, P.A. (1996) *Sea-level Changes.* John Wiley, Chichester.

Pohjola, V.A. and Rogers, J.C. (1997a) Atmospheric circulation and variations in Scandinavian glacier mass balance. *Quaternary Reearch* **47**, 29–36.

Pohjola, V.A. and Rogers, J.C. (1997b) Coupling between the atmospheric circulation and extremes of mass balance of Storglaciären, northern Scandinavia. *Annals of Glaciology* **24**, 229–233.

Porter, P.R., Murray, T. and Dowdeswell, J.A. (1997) Sediment deformation and basal dynamics beneath a glacier surge front: Bakaninbreen, Svalbard. *Annals of Glaciology* **24**, 21–26.

Porter, S.C. (1975) Equilibrium-line altitudes of late Quaternary glaciers in the Southern Alps, New Zealand. *Quaternary Research* **5**, 27–47.

Porter, S.C. (1981a) Recent glacier variations and volcanic eruptions. *Nature* **291**, 139–142.

Porter, S.C. (1981b) Glaciological evidence of Holocene climatic change. In Wigley, T.M.L., Ingram, M.J. and Fermer, C. (eds) *Climate and History.* Cambridge University Press, Cambridge, 82–110.

Porter, S.C. (1986) Pattern and forcing of northern hemisphere glacier variations during the last millennium. *Quaternary Research* **26**, 27–48.

Porter, S.C. (1989) Late Holocene fluctuations of the fiord glacier system in Icy Bay, Alaska, USA. *Arctic and Alpine Research* **21**, 364–379.

Porter, S.C. and Denton, G.H. (1967) Chronology of neoglaciation in the North American Cordillera. *American Journal of Science* **265**, 177–210.

Ram, M. and Koenig, G. (1997) Continuous dust concentration profile of pre-Holocene ice from the Greenland Ice Sheet Project 2 ice core: dust stadials, interstadials, and the Eemian. *Journal of Geophysical Research* **102** (C12), 26641–26648.

Rampino, M.R. and Self, S. (1982) Historic eruptions of Tambora (1815), Krakatau (1883), and Agung (1963), their stratospheric aerosols, and climatic impact. *Quaternary Research* **18**, 127–143.

Raper, S.C.B., Briffa, K.R. and Wigley, T.M.L. (1996) Glacier change in northern Sweden from AD 500: a simple geometric model of Storglaciären. *Journal of Glaciology* **42**, 341–351.

Raymo, M.E., Ruddiman, W.F., Backman, J., Clement, B.M. and Martinson, D.G. (1989) Late Pleistocene evolution of the northern hemisphere ice sheets and North Atlantic deep water circulation. *Paleoceanography* **4**, 413–446.

Raymond, C.F. and Harrison, W.D. (1988) Evolution of Variegated Glacier, Alaska, USA, prior to its surge. *Journal of Glaciology* **34**, 154–169.

Raynaud, D., Barnola, J.M., Chappelaz, J., Zardini, D., Jouzel, J. and Lorius, C. (1992) Glacial–interglacial evolution of greenhouse gases as inferred from ice core analysis: a review of recent results. *Quaternary Science Reviews* **11**, 381–386.

Raynaud, D., Chappellaz, J., Ritz, C. and Martinerie, P. (1997) Air content along the Greenland Ice Core Project core: a record of surface climatic parameters and elevation in central Greenland. *Journal of Geophysical Research* **102** (C12), 26607–26613.

Reeh, N. (1989) Dynamic and climatic history of the Greenland ice sheet. In Fulton, R.J. (ed.) *Quaternary Geology of Canada and Greenland. Geological Surveys of Canada* **1**, 795–822.

Renberg, I. and Segerström, U. (1981) Application of varved lake sediments in palaeoenvironmental studies. *Wahlenbergia* **7**, 125–133.

Reynolds, J.R. and Young, G.J. (1997) Changes in areal extent, elevation and volume of Athabasca Glacier, Alberta, Canada, as estimated from a series of maps produced between 1919 and 1979. *Annals of Glaciology* **24**, 60–65.

Ridley, J.K., Cudlip, W. and Laxon, S.W. (1993) Identification of subglacial lakes using ERS-1 radar altimeter. *Journal of Glaciology* **39**, 625–634.

Rind, D. and Overpeck, J. (1993) Hypothesised causes of decade-to-century-scale climate variability: climate model results. *Quaternary Science Reviews* **12**, 357–374.

Rose, J., Whiteman, C.A., Lee, J., Branch, N.P., Harkness, D.D. and Walden, J. (1997) Mid- and late-Holocene vegetation, surface weathering and glaciation, Fjallsjökull, southeast Iceland. *Holocene* **7**, 457–471.

Ruddiman, W.F. and Kutzbach, J.E. (1990) Late Cenozoic plateau uplift and climate change. *Transactions of the Royal Society of Edinburgh: Earth Sciences* **81**, 301–314.

Ruddiman, W.F. and McIntyre, A. (1981) The North Atlantic Ocean during the last deglaciation. *Palaeogeography, Palaeoclimatology, Palaeoecology* **35**, 145–214.

Ruddiman, W.F., McIntyre, A.F. and Raymo, M.E. (1986) Matuyama 41,000-year cycle: North Atlantic Ocean and northern hemisphere ice sheets. *Earth and Planetary Science Letters* **80**, 117–129.

Ruddiman, W.F. and Raymo, M. (1988) Northern Hemisphere climate regimes during the past 3 Ma: possible tectonic connections. *Philosophical Transactions of the Royal Society, London* **B318**, 411–430.

Ruddiman, W.F., Raymo, M.E., Martinson, D.G., Clement, B.M. and Backman, J. (1989) Pleistocene evolution: Northern Hemisphere ice sheets and North Atlantic Ocean. *Paleoceanography* **4**, 353–412.

Rundgren, M. (1995) Biostratigraphic evidence of the Allerød–Younger Dryas–Preboreal oscillation in northern Iceland. *Quaternary Research* **44**, 405–416.

Röthlisberger, F. (1986) *10000 Jahre Gletschergeschichte der Erde*. Verlag Sauerländer, Aarau.

Saarnisto, M. (1979) Studies of annually laminated sediments. In Berglund, B. (ed.) *Studies of Annually Laminated lake sediments. Palaeohydrological Changes in the Temperate Zone in the Last 15000 Years*. University of Lund, Lund, 61–68.

Salinger, M.J., Heine, M.J. and Burrows, C.J. (1983) Variations of the Stocking (Te Wae Wae) Glacier, Mount Cook, and climatic relationships. *New Zealand Journal of Science* **26**, 321–328.

Saltzman, B. (1985) Paleoclimate modelling. In Hecht, A.D. (ed.) *Paleoclimate Analysis and Modelling*. Wiley, New York, 341–393.

Sarnthein, M., Jansen, E., Weinelt, M., Arnold, M., Duplessy, J.C., Erlenkeuser, H., Flatøy, A., Johannessen, G., Johannessen, T., Jung, S., Koc, N., Labeyrie, L., Maslin, M., Pflaumann, U. and Schulz, H. (1995) Variations in Atlantic surface ocean paleoeceanography, 50°–80°N: a time-slice record of the last 30,000 years. *Paleoceanography* **10**, 1063–1094.

Savoskul, O.S. (1997) Modern and Little Ice Age glaciers in "humid" and "arid" areas of the Tien Shan, central Asia: two different patterns of fluctuations. *Annals of Glaciology* **24**, 142–147.

Schweingruber, F.H. (1988) *Tree Rings: Basics and Applications of Dendrochronology*. Kluwer, Dordrecht.

Schweizer, J. and Iken, A. (1992) The role of bed separation and friction in sliding over an undeformable bed. *Journal of Glaciology* **38**, 77–92.

Schöner, W., Auer, I., Böhm, R., Hammer, N. and Wiesinger, T. (1997) Retreat of Wurtenkees, European East Alps, since 1850. *Annals of Glaciology* **24**, 102–105.

Scuderi, L.A. (1990) Tree-ring evidence for climatically effective volcanic eruptions. *Quaternary Research* **34**, 67–85.

Sear, C.B., Kelly, P.M., Jones, P.D. and Goodess, C.M. (1987) Global surface-temperature responses to major volcanic eruptions. *Nature* **330**, 365–367.

Sejrup, H.-P., Sjøholm, J., Furnes, H., Beyer, I., Eide, L., Jansen, E. and Mangerud, j. (1989) Quaternary tephra chronology on the Iceland Plateau, north of Iceland. *Journal of Quaternary Science* **4**, 109–114.

Self, S., Rampino, M.R. and Barbera, J.J. (1981) The possible effects of large 19th and 20th century volcanic eruptions on zonal and hemispheric surface temperatures. *Journal of Volcanology and Geothermal Research* **11**, 41–60.

Serreze, M.C., Maslanik, J.A., Key, J.R. and Kokaly, R.F. (1995) Diagnosis of the record minimum in Arctic sea ice during 1990 and associated snow cover extremes. *Geophysical Research Letters* **22**, 2183–2186.

Shackleton, N.J. and Pisias, N.G. (1985) Atmospheric carbon dioxide, orbital forcing, and climate. In Sundquist, E.T. and Broecker, W.S. (eds.) *The carbon cycle and atmospheric CO$_2$: natural variations Archean to present. Geophysical Monograph* **32**, 412–417.

Shackleton, N.J. (1987) Oxygen isotopes, ice volume and sea level. *Quaternary Science Reviews* **6**, 183–190.

Shackleton, N.J., Berger, A. and Peltier, W.R. (1990) An alternative astronomical calibration of the lower Pleistocene timescale based on ODP Site 677. *Transactions of the Royal Society of Edinburgh, Earth Sciences* **81**, 251–261.

Shackleton, N.J., Crowhurst, S., Hagelberg, T., Pisias, N.G. and Schneider, D.A. (1995a) A new Late Neogene time scale: application to Leg 138 Sites. In Pisias, N.G., Janacek, L.A., Palmer-Julson, A. and Van Andel, T.H. (eds.) *Proceedings of the Ocean Drilling Program, Scientific Results* **138**, 73–101.

Shackleton, N.J., Hall, M.A. and Pate, D. (1995b) Pliocene stable isotope stratigraphy of Site 846. In Pisias, N.G., Janacek, L.A., Palmer-Julson, A. and Van Andel, T.H. (eds.) *Proceedings of the Ocean Drilling Program, Scientific Results* **138**, 337–355.

Shanaka, L. de S. and Zielinski, G.A. (1998) Global influence of the AD 1600 eruption of Huaynaputina, Peru. *Nature* **393**, 455–458.

Sharp, M. and Dugmore, A. (1985) Holocene glacier fluctuations in eastern Iceland. *Zeitschrift für Gletscherkunde und Glazialgeologie* **21**, 341–349.

Shaw, G.E. (1989) Aerosol transport from sources to ice sheets. In Oeschger, F. and Langway, C.C. (eds) *The Environmental Record in Glaciers and Ice Sheets*. John Wiley, Chichester, 13–28.

Shennan, I. (1989) Holocene crustal movements and sea-level changes in Great Britain. *Journal of Quaternary Science* **4**, 77–89.

Shepherd, M.J. (1987) Glaciation of the tararuas – fact or fiction? In LeHeron, R., Roche, M. and Shepherd, M. (eds) *Geography and Society in a Global Context. Proceedings of the 14th NZ Geography Conference and 56th ANZAAS Congress*. New Zealand Geographical Society, Palmerston North, 114–115.

Shi, Y., Zheng, B. and Li, S. (1992) Last glaciation and maximum glaciation in the Qinghai- Xizang (Tibet) Plateau, a controversy to M. Kuhle's ice sheet hypothesis. *Zeitschrift für Geomorphologie* **Suppl. 84**, 19–35.

Sibrava, V. (1986) Correlation of European glaciations and their relation to the deep-sea record. In Sibrava, V., Bowen, D.Q. and Richmond, G.M. (eds) *Quaternary Glaciations in the Northern Hemisphere. Quaternary Science Reviews* **5**, 433–441.

Sigtryggsson, H. (1972) An outline of sea ice conditions in the vicinity of Iceland. *Jökull* **22**, 1–11.

Sigurdsson, O. (1998) Glacier variations in Iceland 1930–1995. *Jökull* **45**, 3–26.

Sissons, J.B. (1979a) Palaeoclimatic inferences from former glaciers in Scotland and the Lake District. *Nature* **278**, 518–521.

Sissons, J.B. (1979b) The Loch Lomond Stadial in the British Isles. *Nature* **280**, 199–202.

Skinner, B.J. and Porter, S.C. (1987) *Physical Geology*. John Wiley & Sons, Chichester.

Sonett, C.P. and Finney, S.A. (1990) The spectrum of radiocarbon. *Philosophical Transactions of the Royal Society* **A330**, 413–426.

Souchez, R. (1997) The buildup of the ice sheet in central Greenland. *Journal of Geophysical Research* **102** (C12), 26317–26323.

Span, N., Kuhn, M.H. and Schneider, H. (1997) 100 years of ice dynamics of Hintereisferner, central Alps,

Austria, 1894–1994. *Annals of Glaciology* **24**, 297–302.

Stauffer, B., Blunier, T., Dällenbach, A., Indermühle, A., Schwander, J., Stocker, T.F., Tschumo, J., Chappelaz, J., Raynaud, D., Hammer, C.U. and Clausen, H.B. (1998) Atmospheric CO_2 concentration and millennial-scale climate change during the last glacial period. *Nature* **392**, 59–62.

Stauffer, B., Hofer, H., Oeschger, H., Schwander, J. and Siegenthaler, U. (1984) Atmospheric CO_2 concentration during the last glaciation. *Annals of Glaciology* **5**, 160–164.

Steffensen, J.P. (1997) The size distribution of microparticles from selected segments of the Greenland Ice Core Project ice core representing different climatic periods. *Journal of Geophysical Research* **102**, (C12), 26755–26763.

Steig, E.J., Brook, E.J., White, J.W.C., Sucher, C.M., Bender, M.L., Lehman, S.J., Morse, D.L., Waddington, E.D. and Glow, G.D. (1998) Synchronous climate changes in Antarctica and the North Atlantic. *Science* **282**, 92–95.

Street-Perrott, F.A. (1991) General Circulation (GCM) modelling of palaeoclimates: a critique. *Holocene* **1**, 74–80.

Strömberg, B. (1989) Late Weichselian deglaciation and clay varve chronology in east-central Sweden. *Sveriges Geologiska Undersökning* **73**, 1–70.

Stuiver, M. and Brazunias, T.F. (1993) Sun, ocean, climate and atmospheric $^{14}CO_2$: an evaluation of causal and spectral relationships. *Holocene* **3**, 289–305.

Stuiver, M., Brazunias, T.F., Becker, B. and Kromer, B. (1991) Climatic, solar, oceanic and geomagnetic influences on Late-Glacial and Holocene atmospheric $^{14}C/^{12}C$ change. *Quaternary Research* **35**, 1–24.

Stuiver, M., Braziunas, T.F. and Grootes, P.M. (1997) Is there evidence for solar forcing of climate in the GISP2 oxygen isotope record? *Quaternary Research* **48**, 259–266.

Stuiver, M., Denton, G.H., Hughes, T.J. and Fastook, J.L. (1981) History of the marine ice sheet in West Antarctica during the last glaciation: a working hypothesis. In Denton, G.H. and Hughes, T.J. (eds) *The Last Great Ice Sheets*. John Wiley and Sons, New York, 319–439.

Stuiver, M., Grootes, P.M. and Braziunas, T.F. (1995) The GISP2 ^{18}O climate record of the past 16,500 years and the role of the sun, ocean and volcanoes. *Quaternary Research* **44**, 341–354.

Sturm, M., Hall, D.K., Benson, C.S. and Field, W.O. (1991) Non-climatic control of glacier- terminus fluctuations in the Wrangell and Chugach Mountains, Alaska, USA. *Journal of Glaciology* **37**, 348–356.

Stötter, J. (1991) New observations on the postglacial glacial history of Tröllaskagi, northern Iceland. In Maizels, J.K. and Caseldine, C. (eds) *Environmental Change in Iceland: Past and Present*. Kluwer Academic Publishers, Dordrecht, 181–192.

Stötter, J., Wastl, M., Caseldine, C. and Häberle, T. (1999) Holocene palaeoclimatic reconstruction in northern Iceland: approaches and results. *Quaternary Science Reviews* **18**, 457–474.

Sugden, D.E. and John, B.S. (1976) *Glaciers and Landscape*. Edward Arnold, London.

Sundquist, E.T. (1993) The global carbon dioxide budget. *Science* **259**, 934–941.

Sutherland, D.G. (1984) Modern glacier characteristics as a basis for inferring former climates with particular reference to the Loch Lomond Stadial. *Quaternary Science Reviews* **3**, 291–309.

Svendsen, J.I., Astakhov, V.I., Bolshiyanov, D. Y., Demidov, I., Dowdeswell, J.A., Gataullin, V., Hjort, C., Hubberten, H.W., Larsen, E., Mangerud, J., Melles, M., Möller, P., Saarnisto, M. and Siegert, M.J. (1999) Maximum extent of the Eurasian ice sheet in the Barents and Kara Sea region during the Weichselian. *Boreas* **28**, 234–242.

Svendsen, J.I. and Mangerud, J. (1987) Late Weichselian and Holocene sea-level history for a cross section of western Norway. *Journal of Quaternary Science* **2**, 113–132.

Svendsen, J.I. and Mangerud, J. (1992) Paleoclimatic inferences from glacial fluctuations on Svalbard during the last 20 000 years. *Climate Dynamics* **6**, 213–220.

Svendsen, J.I., Mangerud, J., Elverhøi, A., Solheim, A. and Schüttenhelm, R.T.E. (1992) The Late Weichselian glacial maximum on western Spitsbergen inferred from offshore sediments. *Quaternary Science Reviews* **11**, 633–664.

Taylor, K.C., Alley, R.B., Lamorey, G.W. and Mayewski, P. (1997) Electrical measurements on the Greenland Ice Sheet Project 2 Core. *Journal of Geophysical Research* **102**, (C12), 26511–26517.

Teller, J.T. (1995) History and drainage of large ice-dammed lakes along the Laurentide Ice Sheet. *Quaternary International* **28**, 83–92.

Thiede, J. and Bauch, H.A. (1999) The Late Quaternary history of northern Eurasia and the adjacent Arctic Ocean: an introduction to QUEEN. *Boreas* **28**, 3–5.

Thompson Davis, P. (1988) Holocene glacier fluctuations in the American Cordillera. *Quaternary Science Reviews* **7**, 129–157.

Thompson, L.G., Mosley-Thompson, E. and Arnao, B.M. (1984) Major El Nino/Southern Oscillation events recorded in stratigraphy of the tropical Quelccaya Ice Cap. *Science* **226**, 50–52.

Thompson, L.G., *et al.* (1995) Late glacial stage and Holocene tropical ice core records from Huascarán, Peru. *Science* **269**, 46–50.

Thorp, P.W. (1986) A mountain icefield of Loch Lomond Stadial age, western Grampians, Scotland. *Boreas* **15**, 83–97.

Torsnes, I., Rye, N. and Nesje, A. (1993) Modern and Little Ice Age equilibrium-line altitudes on outlet valley glaciers from Jostedalsbreen, western Norway: an evaluation of different approaches to their calculation. *Arctic and Alpine Research* **25**, 106–116.

Tvede, A.M. (1972) *En glasio-klimatisk undersøkelse av Folgefonni*. MSc thesis, University of Oslo.

Tvede, A.M. (1973) Folgefonni, en glasiologisk avviker. *Naturen* **97**, 11–15.

Tvede, A.M. and Liestøl, O. (1977) Blomsterskardbreen, Folgefonni, mass balance and recent fluctuations. *Norsk Polarinstitutt Årbok* **1976**, 225–234.

Tvede, A. and Laumann, T. (1997) Glacier variations on a meso-scale: examples from glaciers in the Aurland Mountains, southern Norway. *Annals of Glaciology* **24**, 130–134.

Tveranger, J., Astakhov, V., Mangerud, J. and Svendsen, J.I. (1999) Surface form of the last Kara ice sheets as inferred from its southwestern marginal features. *Boreas* **28**, 81–91.

UNESCO (1970) Combined heat, ice and water balance at selected glacier basins. *Technical Papers in Hydrology* **5**, 1–20.

Valla, F. and Piedallu, C. (1997) Volumetric variations of Glacier de Sarennes, French Alps, during the last two centuries. *Annals of Glaciology* **24**, 361–366.

van Husen, D. (1997) LGM and Late-Glacial fluctuations in the eastern Alps. *Quaternary International* **38/39**, 109–118.

van der Veen, C.J. (ed.) (1987) *Dynamics of the West Antarctic Ice Sheet*. Kluwer, Dordrecht.

van der Veen, C.J. (1995) Controls on calving rate and basal sliding: observations from Colombia Glacier, Alaska, prior to and during its rapid retreat, 1976–1993. *Byrd Polar Research Center Report 11*.

van der Veen, C.J. (1996) Tidewater calving. *Journal of Glaciology* **40**, 3–15.

van de Wal, R.S.W. and Oerlemans, J. (1995) Response of valley glaciers to climate change and kinematic waves: a study with a numerical ice-flow model. *Journal of Glaciology* **41**, 142–152.

Velichko, A.A., Kononov, Y.M. and Faustova, M.A. (1997) The Last Glaciation on Earth: size and volume of ice sheets. *Quaternary International* **41/42**, 43–51.

Venteris, E.R., Whillans, I.M. and van der Veen, C.J. (1997) Effect of extension rate on terminus position, Columbia Glacier, Alaska, USA. *Annals of Glaciology* **24**, 49–53.

Voloshina, A.P. (1988) Some results of glacier mass balance research on the glaciers of the polar Urals. *Polar Geography and Geology* **12**, 200–211.

Walker, M.J.C. (1995) Climatic changes in Europe during the last glacial-interglacial transition. *Quaternary International* **28**, 63–76.

Walters, R.A. and Meier, M.F. (1989) Variability of glacier mass balances in western North America. In Paterson, D.H. (ed.) *Aspects of Climate Variability in the Pacific and Western Americas*. Geophysical Monograph 55, American Geophysical Union, Washington, DC, 365–374.

Warren, C.R. (1992) Iceberg calving and the glacio-climatic record. *Progress in Physical Geography* **16**, 253–282.

Warren, C.R., Rivera, A. and Post, A. (1997) Greatest Holocene advance of Glacier Pio XI, Chilean Patagonia: possible causes. *Annals of Glaciology* **24**, 11–15.

Warrick, R. and Oerlemans, J. (1990) Sea level rise. In Houghton, J.T., Jenkins, G.J. and Ephraums, J.J. (eds) *Climate Change. IPCC Scientific Assessment*. Cambridge University Press, Cambridge, 257–281.

Weertman, J. (1964) The theory of glacier sliding. *Journal of Glaciology* **5**, 287–303.

Weidick, J.T. (1984) Review of glacier changes in west Greenland. *Zeitschrift für Gletscherkunde und Glazialgeologie* **21**, 301–309.

Weidick, A., Oerter, H. , Reeh, N., Højmark Thomsen, H. and Thorning, L. (1990) The recession of the inland ice margin during the Holocene climatic optimum in the Jakobshavn Isfjord area of west Greenland. *Palaeogeography, Palaeoclimatology, Palaeoecology* **82**, 389–399.

West, R.G., Rose, J., Coxon, P., Ostmaston, H. and Lamb, H.H. (1988) The record of the Cold Stages. In Shackleton, N.J., West, R.G. and Bowen, D.Q. (eds) *The Past Three Million Years: Evolution of Climatic Variability in the North Atlantic Region*, Cambridge University Press, London, 95–112.

Wigley, T.M.L. (1988) The climate of the past 10 000 years and the role of the sun. In Stephenson, F.R. and Wolfendale, A.W. (eds) *Secular Solar and Geomagnetic Variations in the Last 10 000 years*. Kluwer Academic Publishers, Dordrecht, 209–224.

Wigley, T.M.L. and Kelly, P.M. (1990) Holocene climatic change, ^{14}C wiggles and variations in solar irradiance. *Philosophical Transactions of the Royal Society London* **A330**, 547–560.

Wigley, T.M.L. and Raper, S.C.B. (1992) Implications for climate and sea levels of revised IPCC emission scenarios. *Nature* **357**, 293–300.

Wiles, G.C. and Calkin, P.E. (1994) Late Holocene, high-resolution glacial chronologies and climate, Kenai Mountains, Alaska. *Geological Society of America Bulletin* **106**, 281–303.

Williams, R.S., Hall, D.K., Sigurdsson, O. and Chien, J.Y.L. (1997) Comparison of satellite-derived with ground-based measurements of the fluctuations of the margins of Vatnajökull, Iceland, 1973–92. *Annals of Glaciology* **24**, 72–80.

Willis, I.C. (1995) Intra-annual variations in glacier motion: a review. *Progress in Physical Geography* **19**, 61–106.

Winkler, S. (1996) Frührezente und rezente Gletscherstandsschwankungen in Ostalpen und West-/Zentralnorwegen. *Trier Geographische Studien* **15**, 1–580.

Wohlfarth, B. (1996) The chronology of the last termination: a review of radiocarbon-dated, high-resolution terrestrial stratigraphies. *Quaternary Science Reviews* **15**, 267–284.

Wohlfarth, B., Björck, S., Possnert, G., Lemdahl, G., Brunnberg, L., Ising, L., Olsson, S. and Svensson, N.-O. (1993) AMS dating of Swedish varved clays of the last glacial/interglacial transition and the potential/difficulties of calibrating Late Weichselian 'absolute' chronologies. *Boreas* **22**, 113–128.

Wolff, E.W., Moore, J.C., Clausen, H.B. and Hammer, C. (1997) Climatic implications of background acidity and other chemistry derived from electrical studies of the Greenland Ice Core Project ice core. *Journal of Geophysical Research* **102** (C12), 26325–26332.

Woodward, J., Sharp, M. and Arendt, A. (1997) The influence of superimposed-ice formation on the sensitivity of glacier mass balance to climate change. *Annals of Glaciology* **24**, 186–190.

Wright, H.E. Jr, Kutzbach, J.E., Webb, T. III, Ruddiman, W.F., Street-Perrott, F.A. and Bartlein, P.J. (eds) (1993) *Global Climates since the Last Glacial Maximum.* University of Minneapolis Press, Minnesota.

Yakovlev, S.A. (1956) *The Fundamentals of the Quaternary Geology of the Russian Plains.* Trudy, VSEGEI, Leningrad **17** (in Russian).

Yarnal, B. (1984) Relationships between synoptic-scale atmospheric circulation and glacier mass balance in south-western Canada during the International Hydrological Decade, 1965–74. *Journal of Glaciology* **30**, 188–198.

Yen, Y.-C. (1981) *Review of Thermal Properties of Snow, Ice and Sea Ice.* CRREL Report 81-10.

Zheng, B. (1989) Letter to the Editor: Controversy regarding the existence of a large ice sheet on the Qinghai-Xizang (Tibet) Plateau during the Quaternary Period. *Quaternary Research* **32**, 121–123.

Zielinski, G.A. (1998) Determining the range of variability in the volcanism–climate system through multidisciplinary evaluations of explosive eruptions over the last 100,000 years. *PAGES Open Science Meeting London, Abstract of Plenary Lectures*, p.17.

Zielinski, G.A., Mayewski, P.A., Meeker, L.D., Grønvold, K., Germani, M.S., Whitlow, S., Twickler, M.S. and Taylor, K. (1997) Volcanic aerosols records and tephrochronology of the Summit, Greenland, ice cores. *Journal of Geophysical Research* **102** (C12), 26625–26640.

Zreda, M., England, J., Phillips, F., Elmore, D. and Sharma, P. (1999) Unblocking of the Nares Strait by Greenland and Ellesmere ice-sheet retreat 10,000 years ago. *Nature* **398**, 139–142.

Zumbühl, H., Budmiger, G. and Haeberli, W. (1981) Historical documents. In Kasser, P. and Haeberli, W. (eds) *Switzerland and Her Glaciers.* Kümmerly & Frey, Berne, 48–69.

Zuo, Z. and Oerlemans, J. (1997) Numerical modelling of the historic front variations and the future behaviour of the Pasterze glacier, Austria. *Annals of Glaciology* **24**, 234–241.

Zwally, H.J. (1989) Growth of Greenland ice sheet: interpretation. *Science* **246**, 1589–1591.

Østrem, G. (1974) Present alpine ice cover. In Ives, J.D. and Barry R.G. (eds) *Arctic and Alpine Environments.* Methuen, London, 226–250.

Østrem, G. and Olsen, H.Chr. (1987) Sedimentation in a glacier lake. *Geografiska Annaler* **69A**, 123–138.

INDEX

Ablation, 39, 66
Acceleration, 101
Accelerator mass spectrometry (AMS), 137
Accumulation, 66
Accumulation area ratio (AAR), 56, 61, 66
Aerial photogrammetric methods, 56
Aftonian, 120
Airborne scanning laser altimetry, 56
Albedo, 63
Allerød, 135
Analogue experiments, 181
Anglian, 120
Annual moraines, 40
Aphelion, 9
Astronomical theory, 9
Atmospheric pressure, 109

Basal shear stress, 113
Beestonian, 120
Bergschrunds, 107
Biozones, 135
Bølling, 135
Boundary conditions, 181

Calving glacier, 111
Chevron crevasses, 107
Chronozones, 137
Climatically active, 53
COHMAP Members, 181, 182
Cold-based ice, 52
Compressive stress, 101
Continental glaciers, 53
Creep, 103
Crevasses, 103, 107
Critical shear stress, 104
Cromer, 120
Crustal depression, 162
Cryostatic pressure, 109
Cumulative strain, 103

Dansgaard–Oeschger events, 20

Deformable sediments, 50
Devensian, 119, 120
Dielectric profiling (DEP), 26
Dilation, 102
Driving stress, 39
Dump moraines, 40
Dynamic conditions, 181
Dynamically active, 53

Earth Observing System (EOS), 56
Eccentricity of the orbit, 10
Eemian, 119
Elastic strain, 102
Ellipse, 114
Elster, 120
Englacial, 108
Enhanced creep, 105
Equilibrium line altitude, 56, 66
Equipotential surfaces, 109
Erdal event, 142
External forcing mechanism, 6

Firn, 66
Finse event, 142
Foliation, 107
Force, 101
Forcing mechanism, 6
Form drag, 105
Flow model, 39
Fractionation, 18
Fracture, 103
Frictional drag, 106
Frictional heat, 53

Geodetic airborne laser, 56
Geoidal eustasy, 162
Geothermal heat, 53
GISP2, 23
Glacial, 6, 119, 120
Glacier mass balance, 39
Glacio-eustasy, 162

Glacio-isostasy, 162, 164
Global circulation models, 174, 180
Global Environment Monitoring System (GEMS), 3
GRIP, 23
Günz, 120

Heinrich layers, 6
Holstein, 120
Horizontal velocity component, 39
Hoxnian, 120
Hydraulic potential, 108
Hydro-isostasy, 162
Hypsometry, 61, 71

Ice caps, 49
Ice domes, 49
Icefalls, 107
Ice-rafted debris (IRD), 6
Ice sheets, 49
Ice streams, 49
Illinoian, 120
Instantaneous glacierization, 6
Interglacial, 6, 119, 120
Internal deformation, 53
Internal instabilities, 39
International Glacier Commission, 38
Interstadials, 119
Ipswichian, 119
Isobase, 165
Isostasy, 162
Isotope stages, 119

Jökulhlaup, 110

Kansan, 120
Kinematic wave, 39, 106

Lacustrine sediments, 43
Laser altimetry, 56
Latero-frontal fans, 40
Lichenometric dating, 41
Likhvin, 120
Little Ice Age, 144
Loch Lomond, 135
Longitudinal crevasses, 107
Longitudinal stress, 102
Long-wave energy, 64

Marginal crevasses, 107
Marine ice transgression hypothesis, 6
Maritime glaciers, 53
Maunder Minimum, 14, 15
Menap, 120
Mikulino, 120
Milankovitch theory, 9
Mindel, 120
Morozov, 120

Nebraskan, 120
Normal stress, 101
North Atlantic Deep Water (NADW), 121, 161
North Atlantic Oscillation (NAO), 85

Obliquity of the ecliptic, 9
Odessa, 120
Ogive banding, 107
Outlet glaciers, 49
Oxygen isotopes, 11, 121

Palaeoecology, 5
Parabola, 114
PDB, 47
Perihelion, 9
Permanent strain, 102
Polar glaciers, 52
Polythermal glaciers, 52
Porewater pressure, 50
Positive degree days, 69
Precession of the equinoxes, 9
Precession of the solstices, 9
Precipitation elevation gradient, 71
Pressure melting point, 53
Primary permeability, 108
Proxy, 1
Push moraines, 40

Radiocarbon method, 42
Radio-echo sounding, 56
Ramps, 40
Randkluft, 107
Refreezing, 53
Regelation sliding, 105
Regression, 162
Response time, 39, 93
Riss, 119, 120

Saale, 120
Sangamon, 119, 120
Secondary permeability, 108
Sensible heat, 64
Sensitivity experiments, 181
Separation pressure, 106
Séracs, 107
Shear stress, 113
Sheet flow, 50
Short-wave radiation, 63
Sliding, 104
Spaceborne radar, 56
Specific balance, 56
Splaying crevasses, 107
Spörer Mimimum, 15
Stadials, 119
Standard Mean Ocean Water (SMOW), 19, 47
Steady state, 55
Strain, 100, 102

Stream flow, 50
Stress, 100
Subglacial
  meltwater, 108
  sediment deformation, 104
Subpolar glaciers, 52
Sunspots, 14
Superimposed ice zone, 62
Surging glacier, 112
Synthetic aperture radar (SAR), 3

Temperate glaciers, 52
Tensile stress, 102
Thermohaline circulation, 121
Tidewater glaciers, 112
Time lag, 39, 93
Transgression, 162

Transverse crevasses, 107

Valdai, 120, 126

Weichsel, 119, 120
Wet-based ice, 52
Windermere Interstadial, 135
Wisconsin, 119
Wolf Minimum, 15
Wolstonian, 120
World Glacier Monitoring Service (WGMS), 56
Würm, 119

Yarmouth, 120
Yield stress, 102, 104
Younger Dryas, 135

9 780340 706343